INVESTIGATIONS IN NUMBER, DATA, AND SPACE™

3-D Geometry

Exploring Solids and Boxes

Grade 3

Also appropriate for Grade 4

Michael T. Battista
Douglas H. Clements

Developed at TERC, Cambridge, Massachusetts

Dale Seymour Publications

The *Investigations* curriculum was developed at TERC (formerly Technical Education Research Centers) in collaboration with Kent State University and the State University of New York at Buffalo. The work was supported in part by National Science Foundation Grant No. MDR-9050210. TERC is a nonprofit company working to improve mathematics and science education. TERC is located at 2067 Massachusetts Avenue, Cambridge, MA 02140.

This project was supported, in part, by the
National Science Foundation
Opinions expressed are those of the authors and not necessarily those of the Foundation

This book is published by Dale Seymour Publications, an imprint of the Alternative Publishing Group of Addison-Wesley Publishing Company.

Editorial Director: Pat Brill
Project Editor: Priscilla Cox Samii
Series Editor: Beverly Cory
Manuscript Editor: Nancy Tune
ESL Consultant: Nancy Sokol Green
Production/Manufacturing Director: Janet Yearian
Production/Manufacturing Coordinator: Barbara Atmore
Design Manager: Jeff Kelly
Design: Don Taka
Illustrations: DJ Simison, Carl Yoshihara
Cover: Bay Graphics
Composition: Publishing Support Services

Printed in the United States of America.

Printed on Recycled Paper

Order number DS21246
ISBN 0-86651-808-8
3 4 5 6 7 8 9 10-ML-98 97 96 95

Principal Investigator **Susan Jo Russell**
Co-Principal Investigator **Cornelia C. Tierney**
Director of Research and Evaluation **Jan Mokros**

Curriculum Development

Joan Akers
Michael T. Battista
Mary Berle-Carman
Douglas H. Clements
Karen Economopoulos
Ricardo Nemirovsky
Andee Rubin
Susan Jo Russell
Cornelia C. Tierney
Amy Shulman Weinberg

Evaluation and Assessment

Mary Berle-Carman
Abouali Farmanfarmaian
Jan Mokros
Mark Ogonowski
Amy Shulman Weinberg
Tracey Wright
Lisa Yaffee

Teacher Development and Support

Rebecca B. Corwin
Karen Economopoulos
Tracey Wright
Lisa Yaffee

Technology Development

Michael T. Battista
Douglas H. Clements
Julie Sarama Meredith
Andee Rubin

Video Production

David A. Smith

Administration and Production

Amy Catlin
Amy Taber

Cooperating Classrooms for This Unit

Laurie Williams
Hudson Local School District, Hudson, OH

Patsy Bannon
Shaker Heights School District,
Shaker Heights, OH

Jeanne Wall
Arlington Public Schools, Arlington, MA

Consultants and Advisors

Elizabeth Badger
Deborah Lowenberg Ball
Marilyn Burns
Ann Grady
Joanne M. Gurry
James J. Kaput
Steven Leinwand
Mary M. Lindquist
David S. Moore
John Olive
Leslie P. Steffe
Peter Sullivan
Grayson Wheatley
Virginia Woolley
Anne Zarinnia

Graduate Assistants

Kent State University:
Joanne Caniglia, Pam DeLong, Carol King

State University of New York at Buffalo:
Rosa Gonzalez, Sue McMillen,
Julie Sarama Meredith, Sudha Swaminathan

CONTENTS

Teacher Notes

ABOUT THE *INVESTIGATIONS* CURRICULUM

Investigations in Number, Data, and Space is a K–5 mathematics curriculum with four major goals:

- to offer students meaningful mathematical problems
- to emphasize depth in mathematical thinking rather than superficial exposure to a series of fragmented topics
- to communicate mathematics content and pedagogy to teachers
- to substantially expand the pool of mathematically literate students

The *Investigations* curriculum embodies an approach radically different from the traditional textbook-based curriculum. At each grade level, it consists of a set of separate units, each offering 2–4 weeks of work. These units of study are presented through investigations that involve students in the exploration of major mathematical ideas.

Approaching the mathematics content through investigations helps students develop flexibility and confidence in approaching problems, fluency in using mathematical skills and tools to solve problems, and proficiency in evaluating their solutions. Students also build a repertoire of ways to communicate about their mathematical thinking, while their enjoyment and appreciation of mathematics grows.

The investigations are carefully designed to invite all students into mathematics—girls and boys, diverse cultural, ethnic, and language groups, and students with different strengths and interests. Problem contexts often call on students to share experiences from their family, culture, or community. The curriculum eliminates barriers—such as work in isolation from peers, or emphasis on speed and memorization—that exclude some students from participating successfully in mathematics. The following aspects of the curriculum ensure that all students are included in significant mathematics learning:

- Students spend time exploring problems in depth.
- They find more than one solution to many of the problems they work on.
- They invent their own strategies and approaches, rather than relying on memorized procedures.
- They choose from a variety of concrete materials and appropriate technology, including calculators, as a natural part of their everyday mathematical work.
- They express their mathematical thinking through drawing, writing, and talking.
- They work in a variety of groupings—as a whole class, individually, in pairs, and in small groups.
- They move around the classroom as they explore the mathematics in their environment and talk with their peers.

While reading and other language activities are typically given a great deal of time and emphasis in elementary classrooms, mathematics often does not get the time it needs. If students are to experience mathematics in depth, they must have enough time to become engaged in real mathematical problems. We believe that a minimum of five hours of mathematics classroom time a week—about an hour a day—is critical at the elementary level. The plan and pacing of the *Investigations* curriculum is based on that belief.

For further information about the pedagogy and principles that underlie these investigations, see the Teacher Notes throughout the units and the following books:

- *Implementing the* Investigations in Number, Data, and Space™ *Curriculum*
- *Beyond Arithmetic*

The *Investigations* curriculum is presented through a series of teacher books, one for each unit of study. These books not only provide a complete mathematics curriculum for your students, they offer materials to support your own professional development. You, the teacher, are the person who will make this curriculum come alive in the classroom; the book for each unit is your main support system.

While reproducible resources for students are provided, the curriculum does not include student books. Students work actively with objects and experiences in their own environment and with a variety of manipulative materials and technology, rather than with workbooks and worksheets filled with problems. We also make extensive use of the overhead projector as a way to present problems, to focus group discussion, and to help students share ideas and strategies. If an overhead projector is available, we urge you to try it as suggested in the investigations.

Ultimately, every teacher will use these investigations in ways that make sense for his or her particular style, the particular group of students, and the constraints and supports of a particular school environment. We have tried to provide with each unit the best information and guidance for a wide variety of situations, drawn from our collaborations with many teachers and students over many years. Our goal in this book is to help you, as a professional educator, implement this mathematics curriculum in a way that will give all your students access to mathematical power.

Investigation Format

The opening two pages of each investigation help you get ready for the student work that follows. Here you will read:

What Happens—a synopsis of each session or block of sessions.

Mathematical Emphasis—the most important ideas and processes students will encounter in this investigation.

What to Plan Ahead of Time—materials to gather, student sheets to duplicate, transparencies to make, and anything else you need to do before starting.

INVESTIGATION 2

Building Polygons and Polyhedra

What Happens

Sessions 1 and 2: Building Polygons Students build polygons such as triangles, squares, and rectangles. They reflect on the components of the figures and relationships between those components. They discover several properties of triangles, squares, and rectangles.

Session 3: Building Polyhedra Students build polyhedra while looking at pictures or models. They reflect on the parts of the figures and relationships between those parts, gaining a better understanding of the structure of the shapes.

Sessions 4 and 5: Building Polyhedra from Descriptions Students build polyhedra by analyzing verbal descriptions. This requires that they understand the components of these shapes and that they can figure out the structure of the whole figure from its parts.

Mathematical Emphasis

- Recognizing components of polygons—sides and vertices (corners)
- Recognizing how the components of polygons are put together to form whole shapes
- Recognizing the components of polyhedra—the faces, vertices, and edges
- Recognizing how the components of polyhedra are put together to form whole shapes

28 ■ *Investigation 2: Building Polygons and Polyhedra*

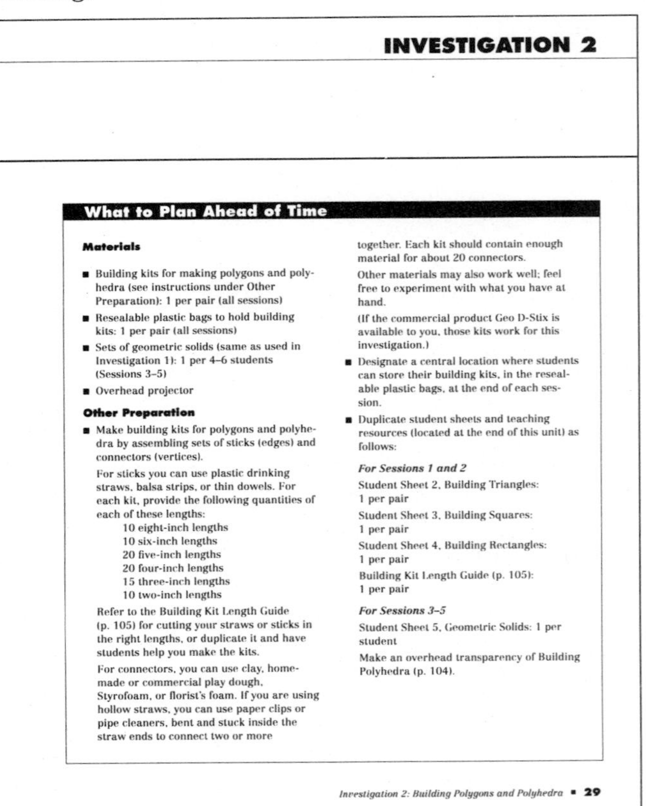

INVESTIGATION 2

What to Plan Ahead of Time

Materials

- Building kits for making polygons and polyhedra (see instructions under Other Preparation): 1 per pair (all sessions)
- Resealable plastic bags to hold building kits: 1 per pair (all sessions)
- Sets of geometric solids (same as used in Investigation 1): 1 per 4–6 students (Sessions 3–5)
- Overhead projector

Other Preparation

- Make building kits for polygons and polyhedra by assembling sets of sticks (edges) and connectors (vertices).

  For sticks you can use plastic drinking straws, balsa strips, or thin dowels. For each kit, provide the following quantities of each of these lengths:

  10 eight-inch lengths
  10 six-inch lengths
  20 five-inch lengths
  20 four-inch lengths
  15 three-inch lengths
  10 two-inch lengths

  Refer to the Building Kit Length Guide (p. 105) for cutting your straws or sticks in the right lengths, or duplicate it and have students help you make the kits.

  For connectors, you can use clay, homemade or commercial play dough, Styrofoam, or florist's foam. If you are using hollow straws, you can use paper clips or pipe cleaners, bent and stuck inside the straw ends to connect two or more together. Each kit should contain enough material for about 20 connectors.

  Other materials may also work well; feel free to experiment with what you have at hand.

  (If the commercial product Geo D-Stix is available to you, those kits work for this investigation.)

- Designate a central location where students can store their building kits, in the resealable plastic bags, at the end of each session.
- Duplicate student sheets and teaching resources (located at the end of this unit) as follows:

  For Sessions 1 and 2
  Student Sheet 2, Building Triangles: 1 per pair
  Student Sheet 3, Building Squares: 1 per pair
  Student Sheet 4, Building Rectangles: 1 per pair
  Building Kit Length Guide (p. 105): 1 per pair

  For Sessions 3–5
  Student Sheet 5, Geometric Solids: 1 per student
  Make an overhead transparency of Building Polyhedra (p. 104).

Investigation 2: Building Polygons and Polyhedra ■ 29

Sessions Within an investigation, the activities are organized by class session, a session being a one-hour math class. Sessions are numbered consecutively through an investigation. Often several sessions are grouped together, presenting a block of activities with a single major focus.

When you find a block of sessions presented together—for example, Sessions 1, 2, and 3—read through the entire block first to understand the overall flow and sequence of the activities. Make some preliminary decisions about how you will divide the activities into three sessions for your class, based on what you know about your students. You may need to modify your initial plans as you progress through the activities, and you may want to make notes in the margins of the pages as reminders for the next time you use the unit.

Be sure to read the Session Follow-Up section at the end of the session block to see what homework assignments and extensions are suggested as you make your initial plans.

While you may be used to a curriculum that tells you exactly what each class session should cover, we have found that the teacher is in a better position to make these decisions. Each unit is flexible and may be handled somewhat differently by every teacher. While we provide guidance for how many sessions a particular group of activities is likely to need, we want you to be active in determining an appropriate pace and the best transition points for your class.

Ten-Minute Math At the beginning of some sessions, you will find Ten-Minute Math activities. These are designed to be used in tandem with the investigations, but not during the math hour. Rather, we hope you will do them whenever you have a spare 10 minutes—maybe before lunch or recess, or at the end of the day.

Ten-Minute Math offers practice in key concepts, but not always those being covered in the unit. For example, in a unit on using data, Ten-Minute Math might revisit geometric activities done earlier in the year. Complete directions for the suggested activities are included at the end of each unit. A compilation of Ten-Minute Math activities is also available as a separate book.

Sessions 4 and 5

Building Polyhedra from Descriptions

Materials

- All materials from Session 3 (geometric solids, building kits, Building Kit Length Guide, and Student Sheet 5)
- Transparency of Building Polyhedra
- Overhead projector

What Happens

Students build polyhedra by analyzing verbal descriptions. This requires that they understand the components of these shapes and that they can figure out the structure of the whole figure from its parts. Their work focuses on:

- determining spatial relationships between the parts of a polyhedron so they can visualize, then build, the whole polyhedron

Activity

Building from Descriptions

Yesterday you built polyhedra by looking at pictures. Today and tomorrow you're going to build polyhedra by listening to descriptions. After we read a description of a polyhedron, try to picture in your mind what it looks like, then build it.

Show the transparency of Building Polyhedra and read the first problem, "Build a polyhedron that has exactly 6 square faces."

Tell students that if they can't figure out the form by picturing it in their mind, they can look at the pictured solids (Student Sheet 5) or the wooden models for help. Doing so, they can count the edges or examine faces of real or pictured objects instead of relying on mental imagery. As students devise solutions, if they seem unsure, ask:

Does your figure fit the description? Are you sure you are correct?

When students have finished the first problem, they compare figures.

Hold up the figures you built. How are these figures the same? How are they different? These two don't look the same. Are both correct? How do you know? How did you figure out how to build this polyhedron?

Students' solution strategies will vary. Some will build the parts of the figure described (for example, some of the square faces), then try to fit the parts together. Others will look for a picture or wooden model that fits the description, then build a figure like it. Still others will simply visualize the solution, then build it from their visual image.

38 ▪ *Investigation 2: Building Polygons and Polyhedra*

Activities The activities include pair and small-group work, individual tasks, and whole-class discussions. We assume that students are seated together, talking and sharing ideas during all work times. Students most often work cooperatively, although each student may record work individually.

Choice Time In some units, some sessions are structured with activity choices. In these cases, students may work simultaneously on different activities focused on the same mathematical ideas. Students choose which activities they want to do, and they cycle through them.

You will need to decide how to set up and introduce these activities and how to let students make their choices. Some teachers present them as station activities, in different parts of the room. Some list the choices on the board as reminders or have students keep their own lists.

Tips for the Linguistically Diverse Classroom At strategic points in each unit, you will find concrete suggestions for simple modifications of the teaching strategies to encourage the participation of all students. Many of these tips offer alternative ways to elicit critical thinking from students at varying levels of English proficiency, as well as from other students who find it difficult to verbalize their thinking.

The tips are supported by suggestions for specific vocabulary work to help ensure that all students can participate fully in the investigations. The Preview for the Linguistically Diverse Classroom (p. 13) lists important words that are assumed as part of the working vocabulary of the unit. Second-language learners will need to become familiar with these words in order to understand the problems and activities they will be doing. These terms can be incorporated into students' second-language work before or during the unit. Activities that can be used to present the words and make them comprehensible are found in the appendix, Vocabulary Support for Second-Language Learners (p. 81).

In addition, ideas for making connections to students' language and cultures, included on the Preview page, help the class explore the unit's concepts from a multicultural perspective.

Session Follow-Up

Homework Homework is not given daily for its own sake, but periodically as it makes sense to have follow-up work at home. Homework may be used for (1) review and practice of work done in class; (2) preparation for activities coming up in class—for example, collecting data for a class project; or (3) involving and informing family members.

Some units in the *Investigations* curriculum have more homework than others, simply because it makes sense for the mathematics that's going on. Other units rely on manipulatives that most students won't have at home, making homework difficult. In any case, homework should always be directly connected to the investigations in the unit, or to work in previous units—never sheets of problems just to keep students busy.

Extensions These follow-up activities are opportunities for some or all students to explore a topic in greater depth or in a different context. They are not designed for "fast" students; mathematics is a multifaceted discipline, and different students will want to go further in different investigations. Look for and encourage the sparks of interest and enthusiasm you see in your students, and use the extensions to help them pursue these interests.

Family Letter A letter that you can send home to students' families is included with the blackline masters for each unit. We want families to be informed about the mathematics work in your classroom; they should be encouraged to participate in and support their children's work. A reminder to send home the letter appears in one of the early investigations. (These letters are also available separately in the following languages: Spanish, Vietnamese, Cantonese, Hmong, and Cambodian.)

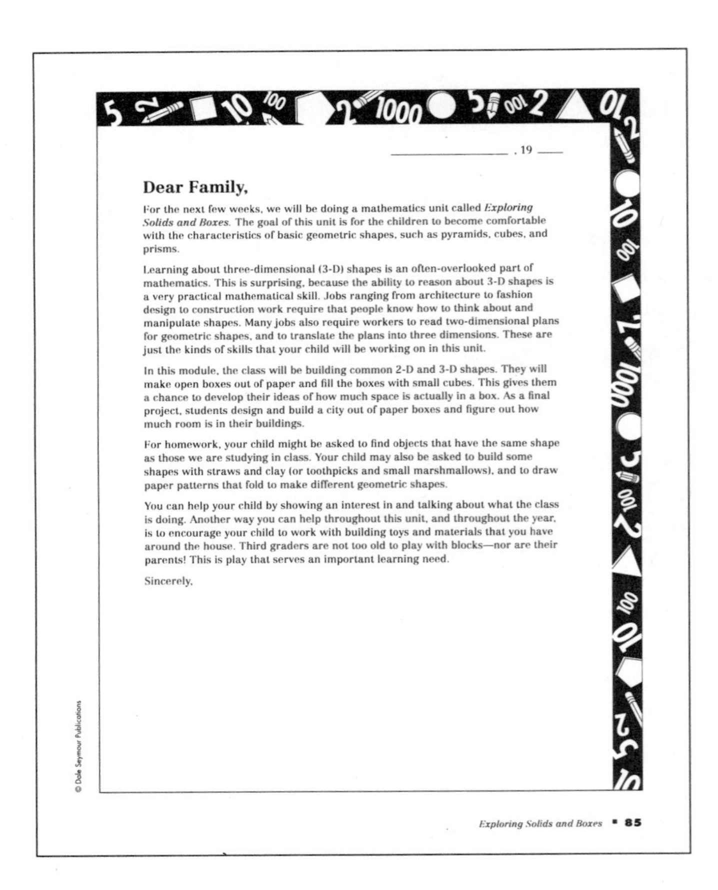

______________, 19 ___

Dear Family,

For the next few weeks, we will be doing a mathematics unit called *Exploring Solids and Boxes*. The goal of this unit is for the children to become comfortable with the characteristics of basic geometric shapes, such as pyramids, cubes, and prisms.

Learning about three-dimensional (3-D) shapes is an often-overlooked part of mathematics. This is surprising, because the ability to reason about 3-D shapes is a very practical mathematical skill. Jobs ranging from architecture to fashion design to construction work require that people know how to think about and manipulate shapes. Many jobs also require workers to read two-dimensional plans for geometric shapes, and to translate the plans into three dimensions. These are just the kinds of skills that your child will be working on in this unit.

In this module, the class will be building common 2-D and 3-D shapes. They will make open boxes out of paper and fill the boxes with small cubes. This gives them a chance to develop their ideas of how much space is actually in a box. As a final project, students design and build a city out of paper boxes and figure out how much room is in their buildings.

For homework, your child might be asked to find objects that have the same shape as those we are studying in class. Your child may also be asked to build some shapes with straws and clay (or toothpicks and small marshmallows), and to draw paper patterns that fold to make different geometric shapes.

You can help your child by showing an interest in and talking about what the class is doing. Another way you can help throughout this unit, and throughout the year, is to encourage your child to work with building toys and materials that you have around the house. Third graders are not too old to play with blocks—nor are their parents! This is play that serves an important learning need.

Sincerely,

Materials

A complete list of the materials needed for the unit is found on p. 10. Some of these materials are available in a kit for the Investigations grade 3 curriculum. Individual items can also be purchased as needed from school supply stores and dealers.

In an active mathematics classroom, certain basic materials should be available at all times: interlocking cubes, pencils, unlined paper, graph paper, calculators, things to count with, and measuring tools. Some activities in this curriculum require scissors and glue sticks or tape. Stick-on notes and large paper are also useful materials throughout.

So that students can independently get what they need at any time, they should know where these materials are kept, how they are stored, and how they are to be returned to the storage area. For example, interlocking cubes are best stored in towers of ten; then, whatever the activity, they should be returned to storage in groups of ten at the end of the hour. You'll find that establishing such routines at the beginning of the year is well worth the time and effort.

Student Sheets and Teaching Resources

Reproducible pages to help you teach the unit are found at the end of this book. These include masters for making overhead transparencies and other teaching tools, as well as student recording sheets.

Many of the field-test teachers requested more sheets to help students record their work, and we have tried to be responsive to this need. At the same time, we think it's important that students find their own ways of organizing and recording their work. They need to learn how to explain their thinking with both drawings and written words, and how to organize their results so someone else can understand them.

To ensure that students get a chance to learn how to represent and organize their own work, we deliberately do not provide student sheets for every activity. We recommend that your students keep a mathematics notebook or folder so that their work, whether on reproducible sheets or their own paper, is always available to them for reference.

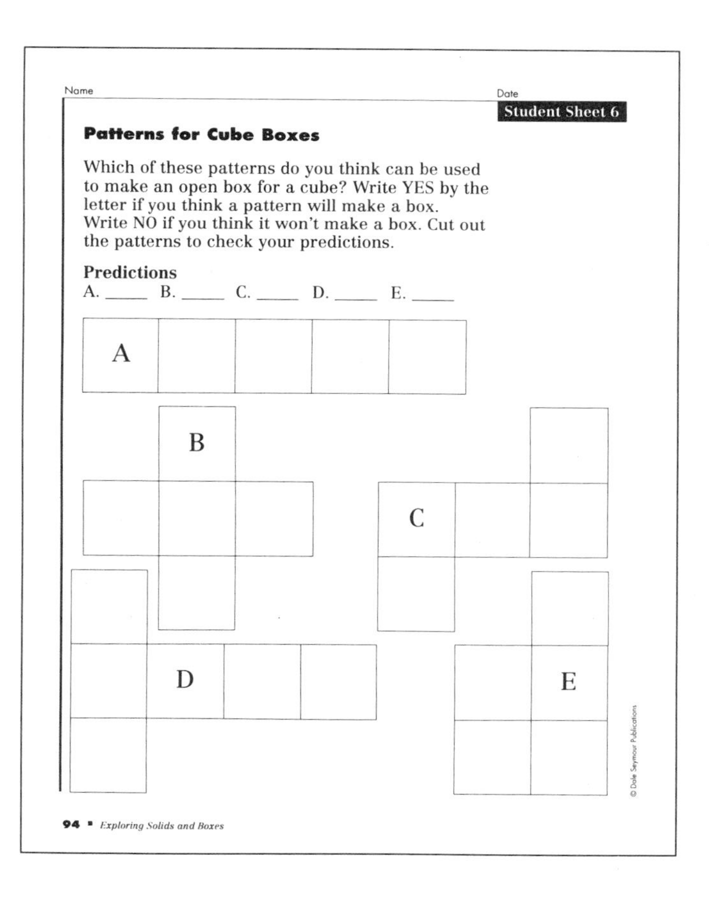

Name Date

Student Sheet 6

Patterns for Cube Boxes

Which of these patterns do you think can be used to make an open box for a cube? Write YES by the letter if you think a pattern will make a box. Write NO if you think it won't make a box. Cut out the patterns to check your predictions.

Predictions
A. _____ B. _____ C. _____ D. _____ E. _____

© Dale Seymour Publications

94 ▪ *Exploring Solids and Boxes*

Help for You, the Teacher

Because we believe strongly that a new curriculum must help teachers think in new ways about mathematics and about their students' mathematical thinking processes, we have included a great deal of material to help you learn more about both.

About the Mathematics in This Unit This introductory section (p. 11) summarizes for you the critical information about the mathematics you will be teaching. This will be particularly valuable to teachers who are accustomed to a traditional textbook-based curriculum.

Teacher Notes These reference notes provide practical information about the mathematics you are teaching and about our experience with how students learn. Many of the notes were written in response to actual questions from teachers, or to discuss important things we saw happening in the field-test classrooms. Some teachers like to read them all before starting the unit, then review them as they come up in particular investigations.

Dialogue Boxes Sample dialogues throughout the unit demonstrate how students typically express their mathematical ideas, what issues and confusions arise in their thinking, and how some teachers have guided class discussions.

These dialogues are based on the extensive classroom testing of this curriculum; many are word-for-word transcriptions of recorded class discussions. They are not always easy reading; sometimes it may take some effort to unravel what the students are trying to say. But this is the value of these dialogues; they offer good clues to how your students may develop and express their approaches and strategies, helping you prepare for your own class discussions.

Where to Start You may not have time to read everything the first time you use this unit. As a first-time user, you will likely focus on understanding the activities and working them out with your students. Read completely through each investigation before starting to present it.

When you next teach this same unit, you can begin to read more of the background. Each time you present this unit, you will learn more about how your students understand the mathematical ideas. The first-time user of *Exploring Solids and Boxes* should read the following:

- About the Mathematics in This Unit (p. 11)
- Teacher Note: Geometric Solids: Types and Terminology (p. 22)

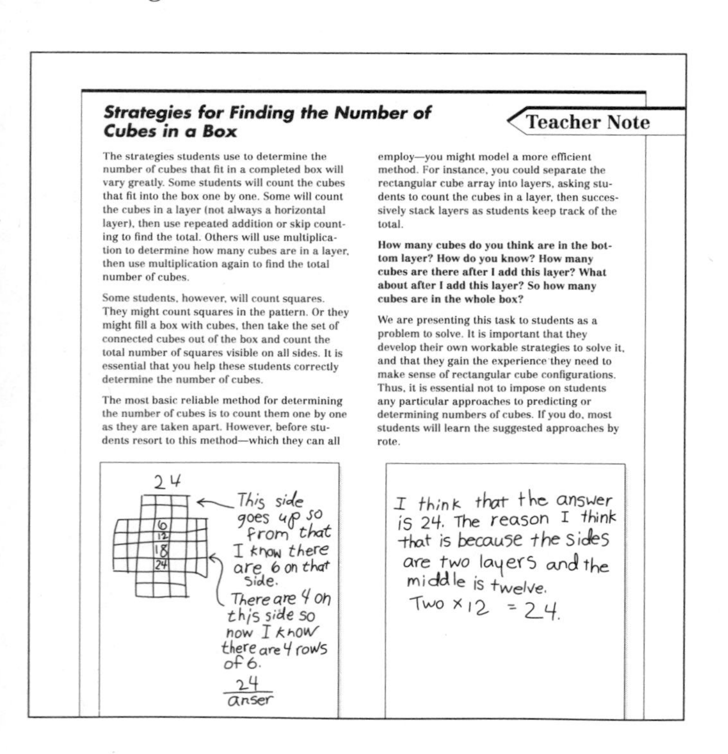

Strategies for Finding the Number of Cubes in a Box — Teacher Note

The strategies students use to determine the number of cubes that fit in a completed box will vary greatly. Some students will count the cubes that fit into the box one by one. Some will count the cubes in a layer (not always a horizontal layer), then use repeated addition or skip counting to find the total. Others will use multiplication to determine how many cubes are in a layer, then use multiplication again to find the total number of cubes.

Some students, however, will count squares. They might count squares in the pattern. Or they might fill a box with cubes, then take the set of connected cubes out of the box and count the total number of squares visible on all sides. It is essential that you help these students correctly determine the number of cubes.

The most basic reliable method for determining the number of cubes is to count them one by one as they are taken apart. However, before students resort to this method—which they can all employ—you might model a more efficient method. For instance, you could separate the rectangular cube array into layers, asking students to count the cubes in a layer, then successively stack layers as students keep track of the total.

How many cubes do you think are in the bottom layer? How do you know? How many cubes are there after I add this layer? What about after I add this layer? So how many cubes are in the whole box?

We are presenting this task to students as a problem to solve. It is important that they develop their own workable strategies to solve it, and that they gain the experience they need to make sense of rectangular cube configurations. Thus, it is essential not to impose on students any particular approaches to predicting or determining numbers of cubes. If you do, most students will learn the suggested approaches by rote.

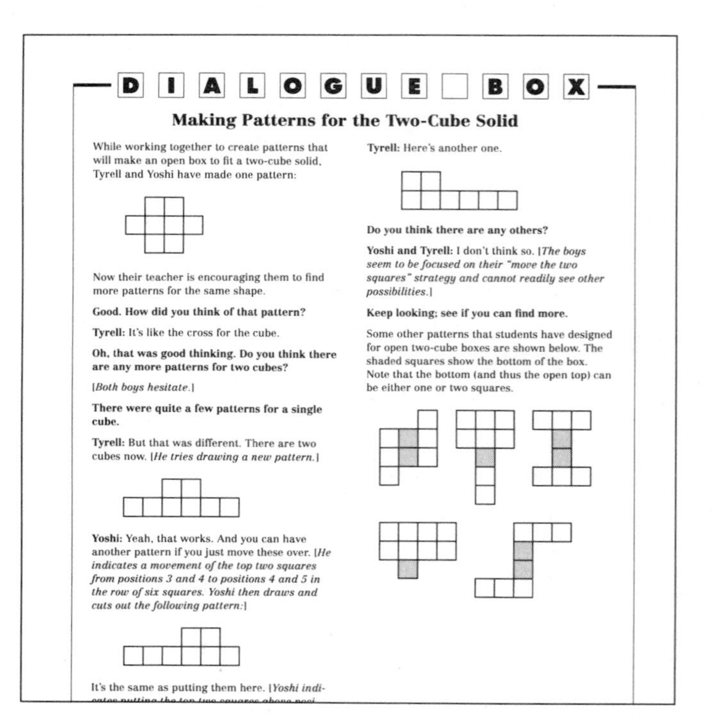

DIALOGUE BOX

Making Patterns for the Two-Cube Solid

While working together to create patterns that will make an open box to fit a two-cube solid, Tyrell and Yoshi have made one pattern:

Now their teacher is encouraging them to find more patterns for the same shape.

Good. How did you think of that pattern?

Tyrell: It's like the cross for the cube.

Oh, that was good thinking. Do you think there are any more patterns for two cubes?

[*Both boys hesitate.*]

There were quite a few patterns for a single cube.

Tyrell: But that was different. There are two cubes now. [*He tries drawing a new pattern.*]

Yoshi: Yeah, that works. And you can have another pattern if you just move these over. [*He indicates a movement of the top two squares from positions 3 and 4 to positions 4 and 5 in the row of six squares. Yoshi then draws and cuts out the following pattern:*]

It's the same as putting them here. [*Yoshi indi-*

Tyrell: Here's another one.

Do you think there are any others?

Yoshi and Tyrell: I don't think so. [*The boys seem to be focused on their "move the two squares" strategy and cannot readily see other possibilities.*]

Keep looking; see if you can find more.

Some other patterns that students have designed for open two-cube boxes are shown below. The shaded squares show the bottom of the box. Note that the bottom (and thus the open top) can be either one or two squares.

Teacher Checkpoints As a teacher of the *Investigations* curriculum, you observe students daily, listen to their discussions, look carefully at their work, and use this information to guide your teaching. We have designated Teacher Checkpoints as natural times to get an overall sense of how your class is doing in the unit.

The Teacher Checkpoints provide a time for you to pause and reflect on your teaching plan while observing students at work in an activity. These sections offer tips on what you should be looking for and how you might adjust your pacing. Are most students fluent with strategies for solving a particular kind of problem? Are they just starting to formulate good strategies? Or are they still struggling with how to start?

Depending on what you see as the students work, you may want to spend more time on similar problems, change some of the problems to use smaller numbers, move quickly to more challenging material, modify subsequent activities for some students, work on particular ideas with a small group, or pair students who have good strategies with those who are having more difficulty.

In *Exploring Solids and Boxes* you will find two Teacher Checkpoints:

Students' Building Strategies (p. 39)
Patterns for Two-Cube Boxes (p. 47)

Embedded Assessment Activities Use the built-in assessments included in this unit to help you examine the work of individual students, figure out what it means, and provide feedback. From the students' point of view, the activities you will be using for assessment are no different from any others; they don't look or feel like traditional tests.

These activities sometimes involve writing and reflecting, at other times a brief interaction between student and teacher, and in still other instances the creation and explanation of a product.

In *Exploring Solids and Boxes* you will find assessment activities in the fourth and fifth investigations:

An 18-Cube Box Pattern (p. 64)
Examining Students' City Plans (p. 73)

Teachers find the hardest part of the assessment to be interpreting their students' work. If you have used a process approach to teaching writing, you will find our mathematics approach familiar. To help with interpretation, we provide guidelines and questions to ask about the students' work. In some cases we include a Teacher Note with specific examples of student work and a commentary on what it indicates. This framework can help you determine how your students are progressing.

As you evaluate students' work, it's important to remember that you're looking for much more than the "right answer." You'll want to know what their strategies are for solving the problem, how well these strategies work, whether they can keep track of and logically organize an approach to the problem, and how they make use of representations and tools to solve the problem.

Ongoing Assessment Good assessment of student work involves a combination of approaches. Some of the things you might do on an ongoing basis include the following:

- **Observation** Circulate around the room to observe students as they work. Watch for the development of their mathematical strategies, and listen to their discussions of mathematical ideas.
- **Portfolios** Ask students to document their work, in journals, notebooks, or portfolios. Periodically review this work to see how their mathematical thinking and writing are changing. Some teachers have students keep a notebook or folder for each unit, while others prefer one mathematics notebook or a portfolio of selected work for the entire year. Take time at the end of each unit for students to choose work for their portfolios. You might also have them write about what they've learned in the unit.

Exploring Solids and Boxes

OVERVIEW

Content of This Unit Students sort, build, and describe different kinds of polygons and common geometric solids so that they become familiar with the components of these shapes and how the components are related. They make paper patterns for solid shapes. They design patterns for rectangular boxes and explore how many unit cubes fit in the boxes. Finally, students build a model of a city consisting of open-box buildings and determine the total amount of room in the buildings.

Because many of the activities in this unit require special materials, it will not always be possible to give related homework assignments. If you want to assign additional homework during this unit, you might use the suggestions given in the Ten-Minute Math sessions.

Connections with Other Units This unit is one of three at grades 3–5 that develop students' knowledge of 3-D geometric objects and their eventual understanding of volume. In another third grade unit (2-D Geometry), *Flips, Turns, and Area,* students also work with polygons. The focus of that unit is seeing shapes in different orientations and covering a flat area with combinations of shapes. The study of polygons is continued in the 2-D Geometry unit, *Turtle Paths,* in which students work on and off the computer to construct squares, rectangles, and triangles.

Work that students do in this unit, when they determine how many cubes will fill a box, is also related to the models for multiplication used in the third grade units on multiplication and division and the number system.

This unit can be used successfully at either grade 3 or grade 4, depending on the previous experience and needs of your students.

Investigations Curriculum ■ Suggested Grade 3 Sequence

Mathematical Thinking at Grade 3 (Introduction)

Things That Come in Groups (Multiplication and Division)

Flips, Turns, and Area (2-D Geometry)

From Paces to Feet (Measuring and Data)

Landmarks in the Hundreds (The Number System)

Up and Down the Number Line (Changes)

Combining and Comparing (Addition and Subtraction)

Turtle Paths (2-D Geometry)

Fair Shares (Fractions)

▶ *Exploring Solids and Boxes* (3-D Geometry)

Investigation 1 • Sorting and Describing Solids

Class Sessions	Activities	Pacing	
Session 1 SORTING POLYHEDRA	Sorting Solids Thinking About Sorting Schemes ■ Homework ■ Extension	1 hr	
Session 2 WHAT'S MY SHAPE?	Learning the Game: What's My Shape? Playing the Game in Small Groups Discussion: How the Game Worked	1 hr	

Investigation 2 • Building Polygons and Polyhedra

Class Sessions	Activities	Pacing	Ten-Minute Math
Sessions 1 and 2 BUILDING POLYGONS	Building the Polygons Discussing the Polygons ■ Homework ■ Extension	2 hrs	Quick Images
Session 3 BUILDING POLYHEDRA	Building the Polyhedra ■ Homework	1 hr	
Sessions 4 and 5 BUILDING POLYHEDRA FROM DESCRIPTIONS	Building from Descriptions ■ Teacher Checkpoint: Students' Building Strategies ■ Homework	2 hrs	

Investigation 3 • Making Boxes

Class Sessions	Activities	Pacing	Ten-Minute Math
Session 1 MAKING BOXES FOR CUBES	Open Boxes for Cubes Cube Patterns More Cube Patterns	1 hr	Quick Images
Session 2 PATTERNS FOR OTHER SOLIDS	■ Teacher Checkpoint: Patterns for Two-Cube Boxes Patterns for Triangular Pyramids ■ Homework ■ Extension	1 hr	

Investigation 4 • How Many Cubes in a Box?

Class Sessions	Activities	Pacing	Ten-Minute Math
Session 1 FINDING THE NUMBER OF CUBES IN A BOX	Predicting the Number of Cubes in a Box Making and Checking Predictions Discussing Student Strategies	1 hr	What Is Likely?
Session 2 TWELVE-CUBE BOXES	Designing Boxes to Hold 12 Cubes ■ Homework	1 hr	
Session 3 PATTERNS FROM THE BOTTOM UP	From Bottom to Sides ■ Assessment: An 18-Cube Box Pattern ■ Homework	1 hr	

Continued on next page

Investigation 5 • Building a City			
Class Sessions	**Activities**	**Pacing**	**Ten-Minute Math**
Sessions 1, 2, 3, and 4 MAKING A BOX CITY	Introducing the Problem Planning and Building the City Final Plans and Construction ■ Assessment: Examining Students' City Plans ■ Extensions	4 hrs	What Is Likely?

MATERIALS LIST

Following are the basic materials needed for the activities in this unit. Items marked with an asterisk are provided with the *Investigations* Materials Kit for grade 3.

* Wooden geometric solids: at least 4 sets per class of 24 students. See page 22 for a diagram of the needed solids.

* For the polygon and polyhedra building kits for each small group, plastic drinking straws cut in the following lengths:

10 eight-inch lengths
10 six-inch lengths
20 five-inch lengths
20 four-inch lengths
15 three-inch lengths
10 two-inch lengths

If possible, each length should be a different color. Alternatively, you might use balsa strips or thin dowels cut in the same lengths.

Connectors for the polygon and polyhedra building kits: material such as clay, homemade or commercial play dough, Styrofoam, or florist's foam (which the straws or sticks can be stuck into). If you are using hollow straws, you can use paper clips or pipe cleaners, bent and stuck inside the straw ends to connect two or more together. Every small group should have enough material to make about 20 connectors.

(The commercial product Geo D-Stix provides connectors and sticks in lengths that will work for this unit.)

Resealable plastic bags to hold building kits

* One-inch cubes: 1 per student

* Interlocking cubes: 60 per student pair, stored in resealable plastic bags or tubs. **Note:** The activities in Investigations 4–5 are designed for cubes that connect on all sides and have edges measuring 3/4 inch. Larger or smaller interlocking cubes with 1 cm or 2 cm edges will work, but corresponding student sheets and teaching resources must be reduced or enlarged to match them. See Investigation 4, What to Plan Ahead of Time (p. 53), for further discussion.

Scissors: 1 per student

Tape: 1 roll per student pair

Overhead projector

The following materials are provided at the end of this unit as blackline masters. They are also available in classroom sets.

Family Letter

Student Sheets 1–10

Teaching Resources:

Building Polyhedra
Building Kit Length Guide
Demonstration Box Pattern
Discussion Box Pattern
One-Inch Graph Paper
Three-Quarter-Inch Graph Paper
Triangle Paper
Quick Image Geometric Designs
Quick Image Cubes

In this unit, students sort and build polygons and common geometric solids so that they become familiar with the components of these shapes and how they are related. They begin to notice and describe important properties of the shapes, for example, how many edges a solid shape has, or how a pyramid has many triangular faces coming to a point. Students develop their own language to describe and compare these shapes while they are exposed to the mathematical terms for the shapes and their components.

As suggested in the NCTM *Standards,* visualization skills and mathematical communication skills are important areas of learning in mathematics and science. One of the major goals of this unit is to promote the development of some basic concepts and language needed to reflect on and communicate about spatial relationships in three-dimensional (3-D) environments.

Throughout the unit, we ask students to use visual imagery to predict solutions to spatial problems before they build the objects. As students build the shapes and compare them to their predictions, they reflect on the differences between what they imagined and what they are building. Through this process, they improve both their visual reasoning and their understanding of the problem situation. We continue to encourage reflection by having students discuss not only their solutions but also the strategies they used to arrive at those solutions.

Because visualization skills have been neglected by traditional mathematics curricula, many students and adults find such tasks difficult. Some teachers have found this unit challenging to teach because they never had experience with these kinds of problems. Many adults remember geometry as the memorization of proofs and theorems; they spent little time handling, observing, constructing, and describing the building blocks of geometry—2-D and 3-D shapes. However, with appropriate experience, everyone can improve his or her ability to visualize—even as an adult!

Through numerous repeated experiences with concrete materials such as those provided in this unit, students' visualization skills will gradually increase. Moreover, many spatial problems that are difficult to visualize are easily solved if you build a concrete model of the situation. So if any of the tasks in this unit seem difficult on first reading, try them using concrete materials, just as the students will be doing.

Because visualization problems involve a set of skills different from those normally taught in mathematics, you may find some students who normally don't do well in mathematics excelling in this unit. These students have high visual ability but not necessarily high language or arithmetic skills. As these students encounter repeated successes, you will see their mathematical self-esteem improve.

During the second half of this unit, students make open boxes from grid paper and investigate the number of cubes that fit inside their boxes. They begin to understand the structure of rectangular boxes and the arrays of cubes that fit inside, for example, that a box that has a 2-unit by 3-unit bottom and is 3 units high can hold three layers of six cubes. Gradually, students learn to visualize the capacity of the boxes without actually filling them with cubes. Understanding the structure of these rectangular prisms is essential for students' later understanding of volume, but it is a difficult concept that will take most students several years and much physical experience to develop.

Many adults solve problems similar to the box tasks in this unit by using a formula: number of cubes = length × width × height. We strongly discourage teaching this formula to students. We have found that most students in elementary school (and even junior high) learn the formula by rote—they have no idea why it works.

In fact, young children's way of approaching these problems does not match the formula. Many of your students will develop an approach based on finding the number of cubes in one layer and adding the layers (or multiplying by the number of layers). As students use skip counting, repeated addition, or multiplication to find the total number of cubes, they are modeling the problems in a powerful and general way and are developing an excellent foundation for later work with volume.

Continued on next page

Mathematical Emphasis At the beginning of each investigation, the Mathematical Emphasis section tells you what is most important for students to learn about during that investigation. Many of these mathematical understandings and processes are difficult and complex. Students gradually learn more and more about each idea over many years of schooling. Individual students will begin and end the unit with different levels of knowledge and skill, but all will develop experience describing and visualizing 2-D and 3-D shapes, gain greater knowledge about relating a 2-D pattern to a 3-D shape, and develop strategies to figure out the number of cubes that fit in a rectangular box.

In the *Investigations* curriculum, mathematical vocabulary is introduced naturally during the activities. We don't ask students to learn definitions of new terms; rather, they come to understand such words as *factor* or *area* or *symmetry* by hearing them used frequently in discussion as they investigate new concepts. This approach is compatible with current theories of second-language acquisition, which emphasize the use of new vocabulary in meaningful contexts while students are actively involved with objects, pictures, and physical movement.

Listed below are some key words used in this unit that will not be new to most English speakers at this age level, but may be unfamiliar to students with limited English proficiency. You will want to spend additional time working on these words with your students who are learning English. If your students are working with a second-language teacher, you might enlist your colleague's aid in familiarizing students with these words, before and during this unit. In the classroom, look for opportunities for students to hear and use these words. Activities you can use to present the words are given in the appendix, Vocabulary Support for Second-Language Learners (p. 81).

overlap When students make patterns that fold into boxes, the patterns cannot have sections that *overlap.*

point, edge Students describe and identify figures by referring to *points* and *edges,* among other features.

Multicultural Extensions for All Students

Whenever possible, encourage students to share words, objects, customs, or any aspects of daily life from their own cultures and backgrounds that are relevant to the activities in this unit. For example:

- As students design and build boxes for packaging in Investigation 3, create a small display of boxes designed for various products. Ask students to bring examples of boxes of products with labels in other languages.
- During Investigation 5, post photographs of cities around the world that students in your class have visited or lived near. You might discuss the shapes of the buildings and the use of space—are buildings tall, using little ground area, or are they more spread out?

Investigations

INVESTIGATION 1

Sorting and Describing Solids

What Happens

Session 1: Sorting Polyhedra Students sort a set of geometric solids several times. Each time, they tell how the solids in each group are alike and how they are different. They discover characteristics of geometric solids, distinguishing between polyhedra and non-polyhedra as well as prisms and pyramids.

Session 2: What's My Shape? Students play a game in which they ask questions in an attempt to identify a mystery solid. As they play, they discover characteristics of different solids. They also develop the language necessary to effectively communicate about solids and their characteristics.

Mathematical Emphasis

- Exploring, sorting, and talking about common geometric solids
- Analyzing how solids are the same and different
- Investigating and analyzing the parts of solids
- Developing the ability to describe solids

What to Plan Ahead of Time

Materials

- Sets of geometric solids that correspond to those on Student Sheet 5, Geometric Solids: 1 per group of 4–6 (all sessions)

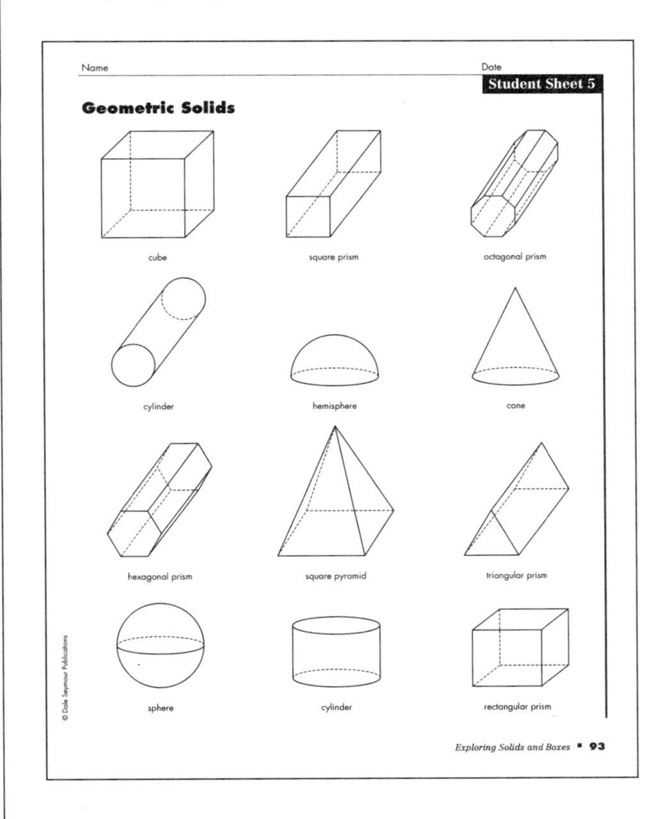

Name Date

Student Sheet 5

Geometric Solids

cube | square prism | octagonal prism

cylinder | hemisphere | cone

hexagonal prism | square pyramid | triangular prism

sphere | cylinder | rectangular prism

© Dale Seymour Publications

Exploring Solids and Boxes ■ 93

Other Preparation

- Use small adhesive labels to number the solids 1–12. For later reference, you might write the number of each solid beside the picture that corresponds to it on the Geometric Solids reference sheet. It is essential that all sets of models be numbered the same; for example, the octagonal prism in each set might be numbered 3. Also, if you mark each set with a different color, sets can be easily regrouped during cleanup time.
- Duplicate the three pages of Student Sheet 1, Identifying Geometric Shapes in the Real World (located at the end of this unit): 1 per student (Session 1, homework)
- Sign the family letter (p. 85) and duplicate to send home with the Session 1 homework.
- If you plan to provide folders in which students will save their work for the entire unit, prepare these for distribution during Session 1.

Session 1

Sorting Polyhedra

Materials

- Sets of geometric solids (1 per group of 4–6)
- Student Sheet 1 (1 set of 3 pages per student, homework)
- Family letter (1 per student)1

What Happens

Students sort a set of geometric solids several times. Each time, they tell how the solids in each group are alike and how they are different. They discover characteristics of geometric solids, distinguishing between polyhedra and non-polyhedra as well as prisms and pyramids. Their work focuses on:

- describing how the solids in each sorted group are alike and different
- attending to the components and properties of different classes of solids, such as polyhedra and non-polyhedra (e.g., cones, cylinders), prisms and pyramids, and so on

Activity

Sorting Solids

Divide the class into as many groups as there are sets of geometric solids (4–6 students per group), and give each group a set. Introduce the sorting problem.

Let's try to figure out how these wooden shapes are the same and how they are different. Put the shapes into groups so that all the shapes in each group are the same in some way.

On a sheet of paper, write the numbers that are on the shapes in each group and circle them to show they go together. Then write how the shapes in each group are the same.

Ask the students to sort the shapes four or five times, including at least one sorting into two groups and one into three groups. Each group of students should keep a recording sheet for their results. For instance, students might sort shapes into two groups and record as follows:

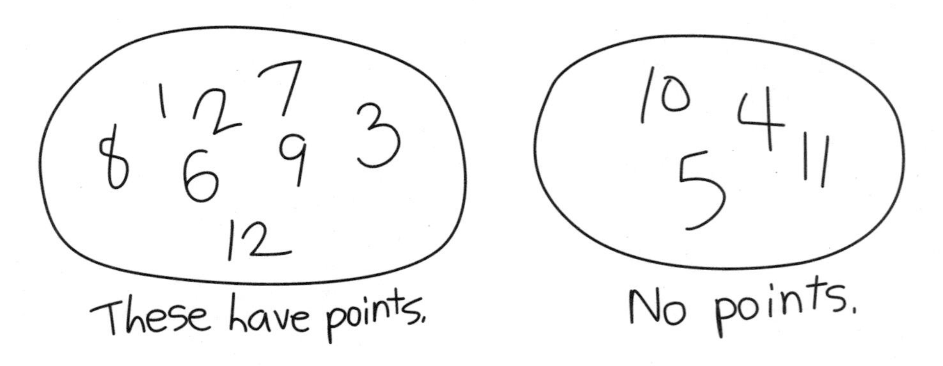

Students will devise numerous idiosyncratic methods of sorting. Many will sort the solids into polyhedra and solids with curved surfaces, describing their categories as "those that roll and those that don't roll." (Some students will include octagonal and hexagonal prisms with the figures that roll.) Some students might sort the figures into shapes with and without points at the top. See the **Teacher Note,** Geometric Solids: Types and Terminology (p. 22), for some simple mathematical distinctions between shapes as well as a discussion of what terminology to use with students.

❖ **Tip for the Linguistically Diverse Classroom** Students with limited English proficiency can contribute to the written descriptions of how the figures in a group are alike by drawing simple pictures over key words. For example:

1 2 7 8 6 9 3 12 — These have points.

10 4 5 11 — No points.

Circulate around the room, asking questions such as the following:

How many groups did you make this time? How are the shapes in this group the same? How are the shapes in these two groups different? Does every shape in this group fit your description? Do any shapes in other groups also fit this description?

See the **Dialogue Box,** Talking to Students About Sorting (p. 23), for another example of how you might discuss this activity with students.

Understanding Students' Approaches Observing students' work and/or collecting their written descriptions of the sorting activities can help you understand what they are attending to. For instance, some students will view the solids only as wholes and will group together only those whose overall shape seems similar: "These are all tall."

Some students will view the solids functionally, saying that the octagonal and hexagonal prisms are alike because "they roll." Some will think of how one shape can be transformed into another, saying, for example, that shapes are alike because "if you take the sides [the flatness] away from these [the octagonal and hexagonal prisms], they will be just like this [the narrow cylinder]." Others will look more at the components of the figures. For example, they might put the square pyramid, the cube, and the narrow square prism together because "they all have squares."

During this session and the next (the What's My Shape? game), you will see students begin to expand the range of shape attributes that they notice.

Activity

Thinking About Sorting Schemes

After students have sorted the figures four or five times, they describe some of their sorting schemes in a whole-class discussion. Guide the discussion by asking the same questions you asked the small groups. Help students focus on their criteria for sorting, that is, on how the shapes in a group are alike, and how they differ from the shapes in another group.

If none of the students have sorted the shapes into polyhedra and non-polyhedra, do so now, placing the cone, sphere, hemisphere, and both cylinders in one group (non-polyhedra) and the rest in the other group (polyhedra). Ask:

How do you think I chose which solids to put into each of these two groups? What do the solids in this group [*point to the polyhedra*] **have in common?**

Some students might characterize them as solids that have lines (edges) and flat surfaces (faces).

If I had another solid, what would it have to be like for me to place it in this group [*point to the polyhedra*]**?**

Now remove the non-polyhedra and sort the remaining solids into prisms and the pyramid. (This will work even better if you can add some other pyramids to the set.)

How do you think I chose which solids to put into each of these two groups? What do the solids in this group [*point to the prisms*] **have in common?**

Some students will say that the prisms do not have points on top. Others might say that the prisms have "tops and bottoms that are the same." Other students might attend to the fact that when the prisms are stood on their bases, the side faces are rectangles, whereas the side faces of a pyramid are triangles.

See the **Dialogue Box,** Talking to Students About Sorting (p. 23), for some examples of students' ideas about the pyramid and prism groups.

❖ **Tip for the Linguistically Diverse Classroom** As students compare and contrast the geometric solids, encourage students to demonstrate their ideas as they say them. For example, if describing a group this way: "They have points, edges, and flat sides," the student shows points and edges and places a hand on a flat side to demonstrate that it is flat. Students with limited English proficiency can use the same demonstration process as you help them articulate their ideas.

Session 1 Follow-Up

Homework

Distribute the three pages of Student Sheet 1, Identifying Geometric Shapes in the Real World. For each picture, suggest that students first write the number of the solid, and check to be sure they have interpreted the pictures correctly.

Students complete the sheet for homework, looking for examples during free time in school, as well as in and around their homes. They can also search in magazines and books, copying or cutting out examples and bringing them to class. Such cutouts can be used to make a bulletin board display. This is a two-day assignment, so students' work can be discussed after Session 2.

Note that some of the polyhedra on the homework sheets may be difficult to find in the "real world." Also, some of the shapes may simply be difficult to recognize. For instance, most pencils are hexagonal prisms, but few students will recognize that until it is pointed out. A stop sign is an octagonal prism, but because it is so thin, most students will not recognize it as such. (Some refuse to accept a stop sign as an example of an octagonal prism even when it is explained.)

Send home the family letter with this homework.

Discover the Sorting Scheme The sorting activity from Session 1 can be extended into a game. Student groups can post one of their sorting solutions on a bulletin board, using circled numbers as they did in class. Other students try to describe the criteria the group used to arrive at this solution.

Teacher Note — Geometric Solids: Types and Terminology

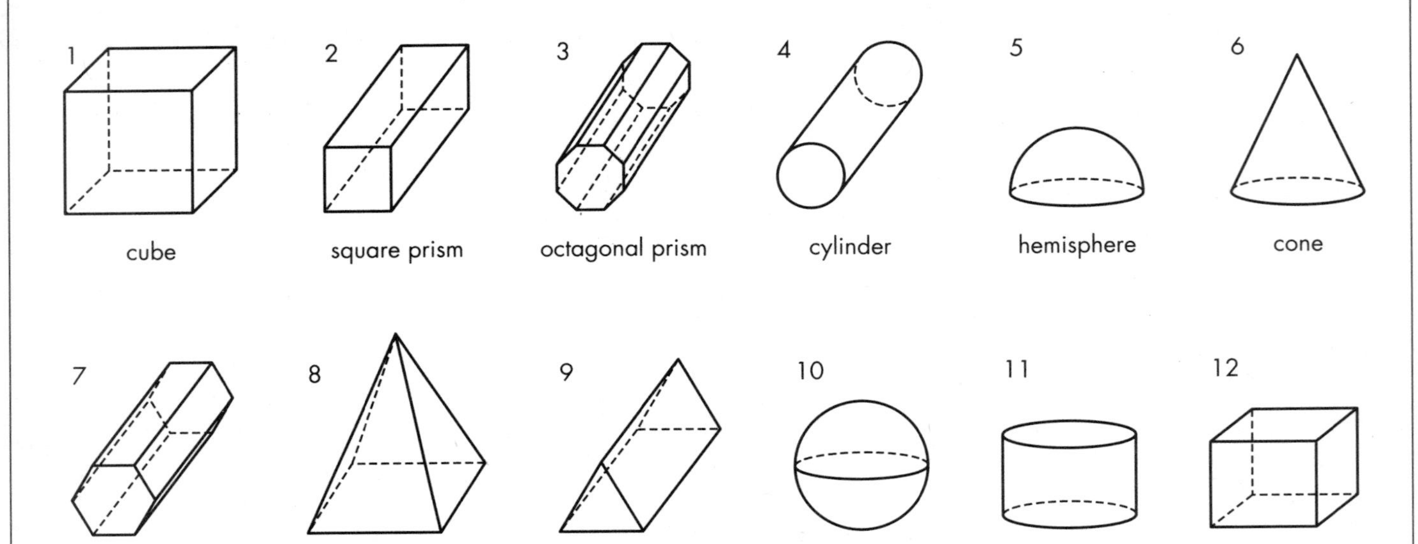

Types of Solids There are many types of geometric solids. Several, such as spheres, cones, and cylinders (figures 4, 5, 6, 10, and 11 above) have some curved surfaces. Others (figures 1, 2, 3, 7, 8, 9, and 12), called *polyhedra,* have only flat surfaces. (*Polyhedron* means having many flat surfaces, or faces.)

Two common types of polyhedra are prisms and pyramids. Students might say that prisms (figures 1, 2, 3, 7, 9, and 12) have "a top and bottom that are the same shape, and all the sides [lateral faces] rectangles." Similarly, they might say that a pyramid (figure 8) has "a flat bottom and a point for a top, with all sides triangles."

Terminology It is not important that students memorize terminology (or acquire abstract definitions) for geometric solids at this grade level. However, you should use the correct terminology (along with terms invented by the students) to familiarize students with it. For instance, you might use the term *polyhedra* or a student-generated term such as *shapes with flat sides* interchangeably, telling students that mathematicians call these shapes polyhedra.

Students will also frequently use *corner* for vertex, *line* for edge, and *side* for face. As long as they are communicating effectively, let them use the language they are comfortable with, while you continue to model the correct language. We will use the terms *edge* and *face,* but will generally use *corner* for vertex, both for polygons and polyhedra. You might, however, wish to use *vertex* [plural *vertices*] interchangeably with *corner.*

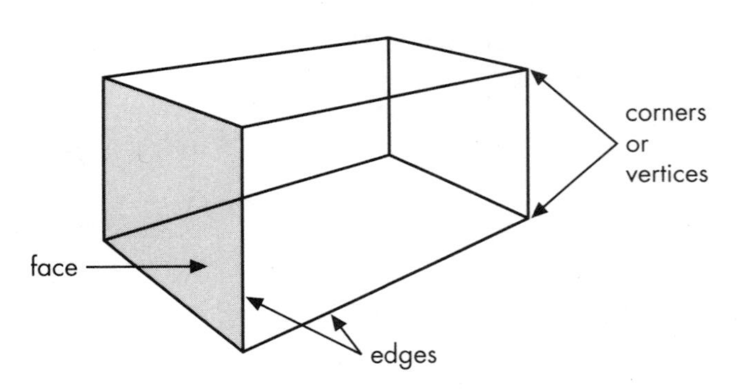

Talking to Students About Sorting

During the class discussion of students' sorting schemes in Session 1, this teacher is talking to the students in a small group about their grouping of polyhedra and non-polyhedra.

How are the shapes in this group [*pointing to the polyhedra*] **the same?**

Samir: The sides are all different. Some are rectangles, here are triangles, and some are squares. These are rectangles even though they are really skinny. [*The teacher notes that this student is focusing on what is different about the shapes in the group rather than what is the same.*]

Annie: They have points, edges, and are flat. This group [*non-polyhedra*] doesn't have them.

What do you mean?

Annie [*referring to a table to illustrate*]**:** This is a point [*touching a corner*]. This is the edge [*touching an edge of the table*]. It's flat [*placing hand down on the top of the table*].

If I had a shape that I wanted to add to this collection, what would it have to look like?

Students: The same thing.

How do you mean?

Jeremy: Points, corners, and sides. Because this one [*cone*] is not in the group. It has a corner and the bottom is flat, but it doesn't have sides that are flat. The sides are round here.

Later in the session, the teacher introduces the idea of sorting the polyhedra only (she has removed the non-polyhedra) into pyramids and prisms.

Now I'm going to sort these [*polyhedra*] **into two groups** [*pyramids and prisms*]. **How are the shapes in the two groups different?**

Liliana: This one [*points to pyramid*], when it goes up it gets thinner. All the other ones [*points to prisms*] stay the same shape.

Do you all agree?

Kate: This one [*points to pyramid*] is like a triangle and these are rectangles.

This figure [*triangular prism*] **is triangular in shape. Why is it not in the group with this one** [*the pyramid*]**?**

Khanh: This [*pyramid*] doesn't have as many sides. This [*triangular prism*] has a top and a bottom. These [*the prisms*] all have both.

Session 2

What's My Shape?

Materials

- Sets of geometric solids (1 per group of 4–6)

What Happens

Students play a game in which they ask questions in an attempt to identify a mystery solid. As they play, they discover characteristics of different solids. They also develop the language necessary to effectively communicate about solids and their characteristics. Their work focuses on:

- attending to the components and properties of different classes of solids, such as polyhedra and non-polyhedra, prisms and pyramids, and so on
- describing how the solids in each group are alike and different

Activity

Learning the Game: What's My Shape?

This identification game will encourage students to make more precise descriptions of geometric solids. The game is first played with the whole class for a couple of rounds, then with small groups. Divide the class into as many groups as there are sets of geometric solids, with 4 to 6 students per group.

To play, one student, the "chooser," picks a shape and writes its number on a sheet of paper—out of sight of the other students. The other students work as a team to determine what the mystery shape is. They take turns asking yes-or-no questions that will help them narrow the possibilities and identify the shape. They must decide as a group on the identity of the mystery shape. Because this is a team effort, students should be encouraged to discuss the implications of their questions, making the thinking process an object of discussion.

For the first round, you should be the chooser. Use this first round to establish the protocol of the game. Encourage students to work cooperatively. Prompt them to clarify ambiguous questions. After each question is answered, students decide together which figures have been eliminated from consideration and move those figures aside.

I've chosen a mystery shape and written its number on this sheet of paper. Your job is to work together and figure out what my mystery shape is.

Questioners in the What's My Shape? game have five rules to follow. Present these rules orally and jot them down on the board as necessary. Students may need to be reminded of these rules while they are learning to play the game.

Rules for Asking Questions

- You must take turns asking questions.
- You can ask only questions that have *yes* or *no* as an answer.
- You cannot ask directly what the shape is (that is, you cannot ask, Is the shape a sphere? Is it number 7?)
- You cannot point to a shape.
- You cannot name an object that looks like the shape (thus, you cannot ask, Does the shape look like a tepee?)

❖ **Tip for the Linguistically Diverse Classroom** Encourage English-proficient students to ensure that their yes-or-no questions are clear to all by using pantomime actions, demonstrations, or quick drawings.

For example, when asking "Does it roll?" the student could use hands to pantomime a rolling movement. Encourage students with limited English proficiency to use similar pantomime, demonstrations, or drawings while you help them phrase their questions.

See the **Dialogue Box,** Playing What's My Shape? (p. 26), for an example of what to expect when first playing the game.

The second time you play as a whole class, select a student to be the chooser, and you can model effective types of questions by asking one or two yourself. We have found that the teacher can, by example, help students learn what is worth noticing about the solids. Students might not think of describing a figure by the number of corners it has, but if a teacher does it once, many students will realize that this is a good way to talk about shapes. (Of course, some students ignore the teacher's example, and that is OK.)

You can also make suggestions when a student is unable to think of a question, for example, "Could you ask a question about the corners or edges of the shapes?"

Activity

Playing the Game in Small Groups

As the students play the game in small groups, continue to watch for communication problems. For instance, some students will consider all vertices as corners, while others will consider only those on the top as corners. Thus, different students might interpret differently the question, Does the shape have four corners? Some students might use *sides* to mean only lateral faces, excluding the tops and bottoms in prisms. Encourage students to explain the meanings of the terms they use and to come to a consensus about those terms.

See the **Dialogue Box,** Building Consensus (p. 27), to see how one class came to agreement on the term *edge*. Consensus building is an important part of mathematics. As a community of people studying geometry, we decide which spatial characteristics or ideas are important and agree on names for them so that we can communicate more effectively.

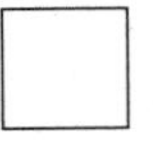

Playing What's My Shape?

Here the teacher is introducing the game of What's My Shape? As the chooser, the teacher is trying to respond to the group's questions with yes-or-no answers, while also trying to help clarify and correct their questioning.

I'm ready for your first question.

Laurie Jo: Does it roll?

Yes ... So, what shapes can it be? Saloni, why don't you put aside the shapes that *won't* work? The rest of you say if you agree with her.

[*Saloni puts all the polyhedra into the box. She also puts the hemisphere away.*]

Laurie Jo: Wait. That one rolls.

Saloni: How?

[*Laurie Jo goes to the front of the room and demonstrates.*]

What do you think? Should this go into the box? [*Most students say no.*] **OK, let's have another question.**

Jamal: Does your shape come to a point?

What do you mean by a point?

Jamal: It's like the top, a sharp part.

OK. No, my shape does not have a point.

[*One student removes the cone and puts it into the box.*]

Cesar: Does it have a circle on the bottom?

Yes. [*The students keep the cylinders and the hemisphere, and remove the remaining figures.*]

Dylan: Does it look like a coffee can?

What can anyone say about the last question?

Jamal: He said what it looked like in real life. You can't do that.

Does anyone have a different question?

Liliana: Is it short and wide?

Discussion: How the Game Worked

Save time at the end of the class period for students to discuss the game.

What problems came up when you were playing the game? Why were those problems?

What were good questions? What questions did not work so well?

Were there any words that you had to agree on? What did you mean by the words *side, corner, point, edge,* [and so on]?

X

Building Consensus

During one game of What's My Shape? a student asks a question about the number of edges of a solid. The teacher notices that not all students are using the term *edge* to mean the same thing and tries to bring this to the students' attention.

[*Showing a triangular prism standing on one of its bases*] **How many edges would you say this has?**

Michael: Three [*he points to the top three edges*].

Do you all agree?

Midori and Amanda: Yeah. Three.

Rashad: Six. Three on top, three on the bottom.

Midori: I change mine to five.

There are many different answers. Could anyone explain to us how you arrived at your answer?

Rashad: I would like to change mine to nine.

OK, Rashad. Could you tell us how you came up with nine?

Rashad: Well [*pointing to each edge*], this one has 1, 2, 3 on the bottom; 4, 5, 6 around the side; 7, 8, 9 on the top.

Could all of you understand Rashad's explanation? [*Most students switch their answers to nine.*]

Michael: I still think it's three. Edges of tables are on top.

Rashad: Edges are sharp. See, the edges on the side and the bottom [*running his fingers along side and bottom edges of the triangular prism*] are sharp, just like the top. They're just like the edges on this desk [*pointing to the teacher's desk*].

What do you think, class? Can we agree on what *edge* means?

Midori: I think Rashad's right, all of them are edges. There's nine.

Almost all students, including Michael, now agree.

INVESTIGATION 2

Building Polygons and Polyhedra

What Happens

Sessions 1 and 2: Building Polygons Students build polygons such as triangles, squares, and rectangles. They reflect on the components of the figures and relationships between those components. They discover several properties of triangles, squares, and rectangles.

Session 3: Building Polyhedra Students build polyhedra while looking at pictures or models. They reflect on the parts of the figures and relationships between those parts, gaining a better understanding of the structure of the shapes.

Sessions 4 and 5: Building Polyhedra from Descriptions Students build polyhedra by analyzing verbal descriptions. This requires that they understand the components of these shapes and that they can figure out the structure of the whole figure from its parts.

Mathematical Emphasis

- Recognizing components of polygons—sides and vertices (corners)
- Recognizing how the components of polygons are put together to form whole shapes
- Recognizing the components of polyhedra—the faces, vertices, and edges
- Recognizing how the components of polyhedra are put together to form whole shapes

What to Plan Ahead of Time

Materials

- Building kits for making polygons and polyhedra (see instructions under Other Preparation): 1 per pair (all sessions)
- Resealable plastic bags to hold building kits: 1 per pair (all sessions)
- Sets of geometric solids (same as used in Investigation 1): 1 per 4–6 students (Sessions 3–5)
- Overhead projector

Other Preparation

- Make building kits for polygons and polyhedra by assembling sets of sticks (edges) and connectors (vertices).

  For sticks you can use plastic drinking straws, balsa strips, or thin dowels. For each kit, provide the following quantities of each of these lengths:

  10 eight-inch lengths
  10 six-inch lengths
  20 five-inch lengths
  20 four-inch lengths
  15 three-inch lengths
  10 two-inch lengths

  Refer to the Building Kit Length Guide (p. 105) for cutting your straws or sticks in the right lengths, or duplicate it and have students help you make the kits.

  For connectors, you can use clay, homemade or commercial play dough, Styrofoam, or florist's foam. If you are using hollow straws, you can use paper clips or pipe cleaners, bent and stuck inside the straw ends to connect two or more together. Each kit should contain enough material for about 20 connectors.

  Other materials may also work well; feel free to experiment with what you have at hand.

  (If the commercial product Geo D-Stix is available to you, those kits work for this investigation.)

- Designate a central location where students can store their building kits, in the resealable plastic bags, at the end of each session.
- Duplicate student sheets and teaching resources (located at the end of this unit) as follows:

  For Sessions 1 and 2

  Student Sheet 2, Building Triangles: 1 per pair

  Student Sheet 3, Building Squares: 1 per pair

  Student Sheet 4, Building Rectangles: 1 per pair

  Building Kit Length Guide (p. 105): 1 per pair

  For Sessions 3–5

  Student Sheet 5, Geometric Solids: 1 per student

  Make an overhead transparency of Building Polyhedra (p. 104).

Sessions 1 and 2

Building Polygons

Materials

- Building kits: 1 per small group
- Building Kit Length Guide: 1 per pair
- Student Sheets 2–4 (1 each per pair)
- Overhead projector

What Happens

Students build polygons such as triangles, squares, and rectangles. They reflect on the components of the figures and relationships between those components. They discover several properties of triangles, squares, and rectangles. Their work focuses on:

- finding the number of sides and corners in polygons
- investigating relationships between lengths of sides of polygons

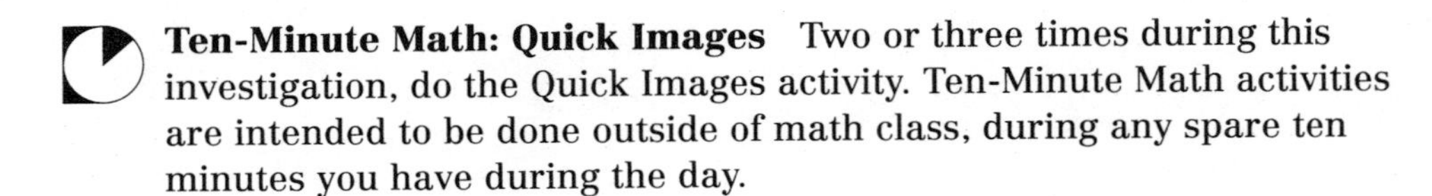

Ten-Minute Math: Quick Images Two or three times during this investigation, do the Quick Images activity. Ten-Minute Math activities are intended to be done outside of math class, during any spare ten minutes you have during the day.

Choose a design cut from the Quick Image Geometric Designs transparency (p. 111). Students will need only paper and pencil. Show the design on the overhead projector for 3 seconds. Students try to draw the image and figure out how it is put together.

Show the design for another 3 seconds, and let students revise their drawings. Then reveal the design for final comparisons. Ask students to describe how they saw the image on successive flashes.

For full directions and variations, see pp. 77–78.

Activity

Building the Polygons

We're going to build some familiar shapes such as triangles, squares, and rectangles. These shapes are called *polygons*.

As you introduce this activity, draw some examples on the board.

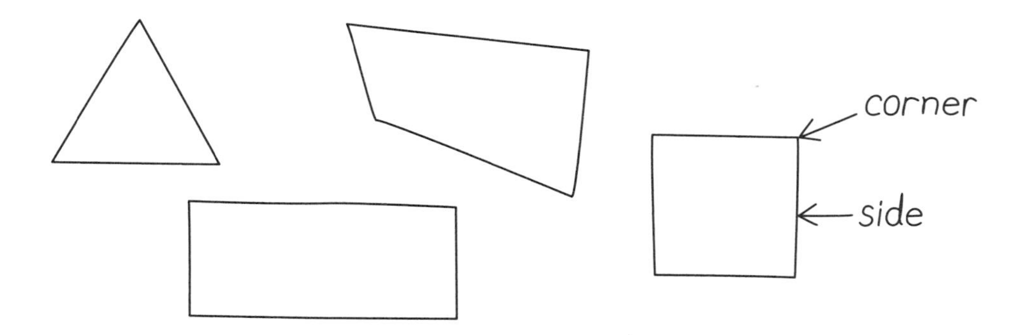

We'll call the straight parts of a polygon *sides* and the "pointy parts" where the sides come together *corners*. [*Point to a few examples of sides and corners.*] **Who can show me a side and a corner for this rectangle?**

Distribute one set of building materials, a Building Kit Length Guide, and Student Sheets 2 and 3, Building Triangles and Building Squares, to each small group of students. Students will share these materials throughout this investigation.

Demonstrate how to build a polygon with the sticks and connectors you have assembled for your students' kits.

Over these two sessions, students build the polygons and solve the problems on the sheets. Generally, they will have to take apart one figure before building another. Students draw each polygon that they make, labeling the sides appropriately. See the **Teacher Note,** Naming Sticks (p. 34), for a discussion of how students can refer to the different lengths of sticks as they talk and write about them.

It is important to have a whole-class discussion of student work on the same day students complete a given student sheet; see the next activity for discussion suggestions. It usually takes about one session for students to complete both Student Sheet 2 (the triangles) and Student Sheet 3 (the squares), allowing for a 15-minute follow-up class discussion. The class can then move on to Student Sheet 4, Building Rectangles, with a follow-up discussion of that work as they finish it.

The challenge problems on Student Sheets 2–4 make a good extension activity; see discussion on p. 33.

Activity

Discussing the Polygons

Hold the appropriate section of this discussion as students finish their work on the three student sheets. You might be holding a separate discussion for each type of figure.

Triangles

In discussing Student Sheet 2, Building Triangles, encourage students to draw on the overhead or hold up to show their constructed triangles.

For problem 1, building triangles with all three edges the same length, students should notice that although their triangles may be different sizes, all are the same shape. Place two such similar but different-sized triangles on top of one another to emphasize this. As one student observed, "The corners are the same—the sides are just bigger."

Then discuss problem 3:

Show me three sticks that did not make a triangle. Why don't these sticks make a triangle?

If one stick is longer than the sum of the lengths of the other two, a triangle cannot be made. Few, if any, students will be able to express this idea concisely. The best description that one class of students could devise was, "You can't have two short ones [sticks] and one that's much longer."

Even though students have found sticks that don't make a triangle, they might not realize the implications of their findings. Thus it is worth asking:

Will *every* three sticks make a triangle?

Squares

As students complete their work on Student Sheet 3, Building Squares, ask questions that will help them focus on the properties of squares.

How many sticks did you use for each square?
What did you notice about the lengths of the sides of a square?
Can you build a square with one 6-inch, one 5-inch, one 4-inch, and one 3-inch stick? Why or why not?

Students should have discovered that a square can be made only by using four sticks of the same length.

❖ **Tip for the Linguistically Diverse Classroom** Students with limited English proficiency could draw or demonstrate to you their answers to item 2 on both Student Sheets 3 and 4.

Rectangles

In your discussion of their work on Student Sheet 4, Building Rectangles, help students focus on the properties of rectangles with questions like these:

How many sides are there in each rectangle? What do you notice about the lengths of the sides of these rectangles?
Can you build a rectangle with two 3-inch sticks, one 5-inch stick, and one 6-inch stick? Why or why not?

Students should discover that a rectangle can be made only from two pairs of same-length sticks.

Sessions 1 and 2 Follow-Up

If you think your students can obtain the necessary materials, they can build polygons at home using toothpicks or straws for sticks and small marshmallows, clay, or jelly beans for fasteners.

Polygon Challenges For the challenge problem on Student Sheet 2, there is only one triangle that can be built using one 5-inch, one 6-inch, and one 8-inch stick, although there are many different orientations for this triangle. For the challenge problem on Student Sheet 4, there are an infinite number of four-sided polygons that can be built by using one 6-inch, one 5-inch, one 8-inch, and one 4-inch stick. Different four-sided polygons can be obtained by changing the order of the sticks. But even with a given order of sticks, an infinite number of different shapes can be obtained by simply changing the angles of the shape. (This cannot be done with a triangle. That is why a triangle is called a "rigid figure," and why triangles are used to brace many structures.)

An issue that often comes up with these activities is how to decide whether two shapes are the same. Mathematicians consider two figures to be the same if they have exactly the same shape, no matter how they are oriented. Many students do not accept this convention, saying that two copies of the same figure are different if they are oriented differently. The **Dialogue Box,** Are They the Same? (p. 35) provides an example of how one teacher handled this problem.

Teacher Note

Naming Sticks

Once you have assembled the students' building kits, you will need to establish a way of referring to the different sticks. If your sticks are in different colors, they can be referred to by color name; if not, they can be named by their length. For example, you might say to students, "Get one 5-inch stick and two 6-inch sticks." Students can use their Building Kit Length Guide to find these sticks quickly. They can simply compare a stick to the lines on the guide to determine its length.

Note, however, that even though you may be referring to the sticks by their length, many students will think of the numbers merely as labels, not measurements. That is, for many students, *5-inch stick* will be understood much the same way as *red stick;* they won't think of it as a stick that measures 5 inches. This is because many students at this age have not developed a firm grasp of measurement.

Students' use of such a "labeling" conception of measures is acceptable for the purposes of this unit. In fact, the activities in this investigation can actually help students take the first steps toward seeing how the measurement of length is relevant in the study of geometric shapes.

Are They the Same?

These two students are working together on the challenge problem on Student Sheet 4: How many different four-sided polygons can you build using one 6-inch stick, one 5-inch stick, one 8-inch stick, and one 4-inch stick?

Annie: There's only one.

Are you sure?

Latisha: Well, if you flip it over, it's still the same.

What if you rearrange the sides?

Annie: Oh yeah. Here's another.

Latisha: How do we know there's not more? I want to keep trying.

Annie: How do you know they're different? Doesn't it equal one we already made?

Latisha [*comparing one figure to her picture of a previously built polygon*]: No. This has a longer top.

Why don't you leave them together after you make them? Then when you make another, you could compare.

Annie [*making another example*]: These two [*the two figures shown below*] are definitely not the same.

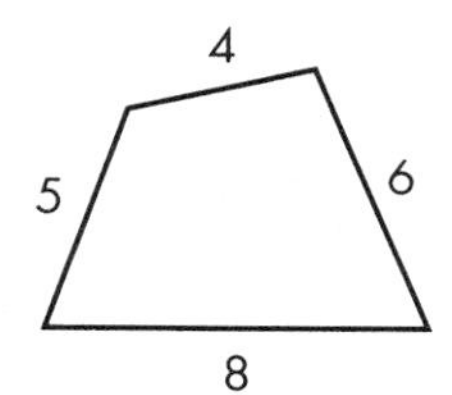

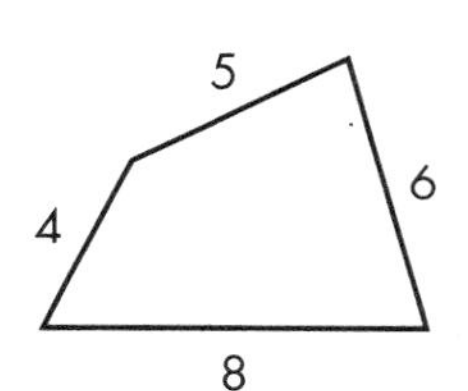

Latisha: Yes they are. They look the same to me.

Annie: Prove it to me.

Latisha [*after examining the figures more carefully*]: Well, I guess they're different. The shortest and longest sticks are connected in this one [*points to figure on right*], and they aren't in the other one.

Annie: Look, I can't make this one fit exactly on top of that one no matter how I turn it ... even if I flip it.

These students have dealt with two difficulties that arise with this problem. First, they had to keep track of the figures they had already made. Apparently their drawings were not accurate enough to distinguish one figure from another. So, at the teacher's suggestion, they compared actual figures rather than drawn ones.

Second, they had to go beyond comparing the overall shapes of the figures. They did this by looking at the parts—the sides—and by trying to superimpose one figure onto another, an intuitive method of judging congruence. They did not, however, recognize that different shapes could be obtained by changing the angles of the shapes they made.

Session 3

Building Polyhedra

Materials

- Geometric solids (1 set per group of 4–6)
- Building kits and Building Kit Length Guide from Sessions 1–2 (1 per pair)
- Student Sheet 5 (1 per student)

What Happens

Students build polyhedra while looking at pictures or models. They reflect on the parts of the figures and relationships between those parts, gaining a better understanding of the structure of the shapes. Their work focuses on:

- determining the spatial relationships between the faces, edges, and corners that are necessary for the polyhedra to be built
- organizing their conceptions of polyhedra so they can count the faces, edges, and corners

Ten-Minute Math: Quick Images Continue to do the Quick Images activity with Geometric Designs over the next few days. You may want to prepare some of your own simple designs that incorporate some of the 2-D shapes students have been making.

Activity

Building the Polyhedra

Each pair of students should have a building kit and a Building Kit Length Guide to share (the same ones they used in Sessions 1–2).

Distribute Student Sheet 5, Geometric Solids. (Students will use this sheet during Sessions 4 and 5, as well.) Also distribute the sets of geometric solids; 2 or 3 student pairs can share each set.

Which figures on this sheet can you build with the sticks and connectors? There is no bending of sticks allowed. If you can't figure out how to build a figure by looking at its picture, you can look at the wooden model to help.

Beside each picture of a figure that you can build, write how many edges and corners the figure has. If you cannot build a figure with your kit, write *No* beside it.

At this grade level, counting the parts of a polyhedron is not a trivial activity. Some students miscount parts—such as the edges—because they do not conceptually organize what is being counted, and they lose track.

As you work with pairs of students, or even in the whole-class discussion, you might demonstrate organized thinking during counting. For a rectangular prism, for example, you might say, "I counted four edges on the bottom, four more on the top, and four more on the sides."

After students have completed the building activity, hold a class discussion about their work.

Which figures could you build with your sticks and connectors? How are these figures different from the ones that you could not build?

During the discussion, encourage students to use the geometric names that appear on Student Sheet 5, but allow them to use more informal names, such as *box*. Students are not expected to memorize the geometric names. You may want to refer back to the **Teacher Note,** Geometric Solids: Types and Terminology (p. 22), for this discussion.

Once students have characterized the figures that can be built with the kits (saying perhaps that they have only flat sides), you can recall for them that these figures are called polyhedra. The word *polyhedron* (or its plural *polyhedra*) means "having many faces."

Pointing first to some solid polyhedra, then to a couple of polyhedra that the students have built, ask how many faces the polyhedra have and what the shapes of these faces are.

Session 3 Follow-Up

Homework

If you think your students can obtain the necessary materials, they might build polyhedra at home using toothpicks or straws for sticks and small marshmallows, clay, or jelly beans for connectors.

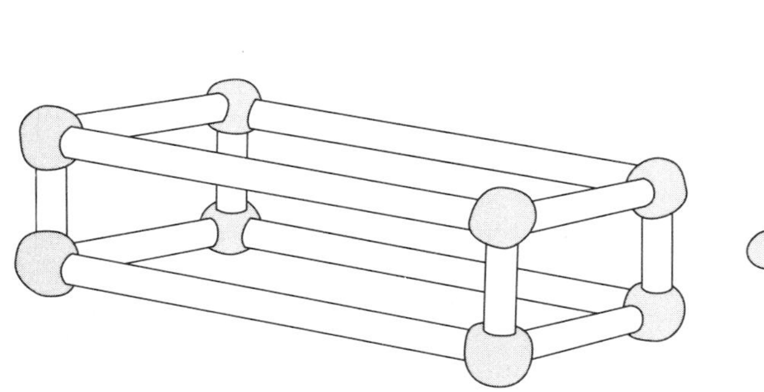

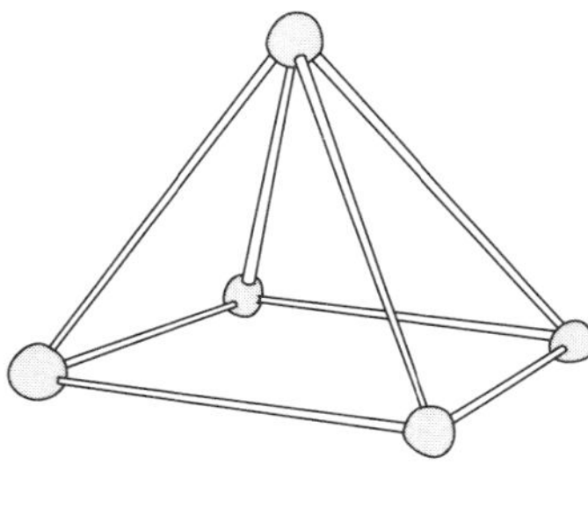

Sessions 4 and 5

Building Polyhedra from Descriptions

Materials

- All materials from Session 3 (geometric solids, building kits, Building Kit Length Guide, and Student Sheet 5)
- Transparency of Building Polyhedra
- Overhead projector

What Happens

Students build polyhedra by analyzing verbal descriptions. This requires that they understand the components of these shapes and that they can figure out the structure of the whole figure from its parts. Their work focuses on:

- determining spatial relationships between the parts of a polyhedron so they can visualize, then build, the whole polyhedron

Activity

Building from Descriptions

Yesterday you built polyhedra by looking at pictures. Today and tomorrow you're going to build polyhedra by listening to descriptions. After we read a description of a polyhedron, try to picture in your mind what it looks like, then build it.

Show the transparency of Building Polyhedra and read the first problem, "Build a polyhedron that has exactly 6 square faces."

Tell students that if they can't figure out the form by picturing it in their mind, they can look at the pictured solids (Student Sheet 5) or the wooden models for help. Doing so, they can count the edges or examine faces of real or pictured objects instead of relying on mental imagery. As students devise solutions, if they seem unsure, ask:

Does your figure fit the description? Are you sure you are correct?

When students have finished the first problem, they compare figures.

Hold up the figures you built. How are these figures the same? How are they different? These two don't look the same. Are both correct? How do you know? How did you figure out how to build this polyhedron?

Students' solution strategies will vary. Some will build the parts of the figure described (for example, some of the square faces), then try to fit the parts together. Others will look for a picture or wooden model that fits the description, then build a figure like it. Still others will simply visualize the solution, then build it from their visual image.

Students then build the remaining figures described on the transparency, discussing each as a class when they have finished building. (Students should disassemble one figure before building the next.)

To answer problem 6, students should name the figures they make, tell how many corners and edges they have (and the lengths of the sticks used), and describe the shapes of the faces. The emphasis in this activity is on using imagery to generate ideas, not exhausting all the possibilities.

Solutions for the problems on the Building Polyhedra transparency are as follows: (1) any cube; (2) any square pyramid; (3) any triangular prism; (4) any rectangular prism, including a cube; (5) any square pyramid; (6) any rectangular prism or hexagonal pyramid—there are other less familiar possibilities; (Challenge problem) any triangular pyramid.

Teacher Checkpoint

Students' Building Strategies

The preceding activity provides information on the progress students have made in understanding the structure of common polyhedra.

- Students who cannot build the polyhedra from verbal descriptions, even with the pictures on Student Sheet 5 (Geometric Solids) and the wooden models to refer to, may need a review of polyhedra, corners, faces, and edges to be sure they understand the concepts. Some students may simply need more experience building polyhedra directly from the wooden models or from diagrams.
- Students who must use the reference sheet or models, but who can correctly make most of the polyhedra, have developed practical knowledge of the concept of polyhedra and their components—corners, faces, and edges.
- Students who can complete the problems without looking at the reference sheet or models have developed not only excellent visualization skills, but a more sophisticated understanding of polyhedra and relationships between their components.

Sessions 4 and 5 Follow-Up

Students might continue to build polyhedra at home using toothpicks or straws for sticks and small marshmallows, clay, or jelly beans for fasteners. They might also try creating their own verbal description problems for the polyhedra they build; they could trade these problems with classmates for more work in building from written descriptions.

INVESTIGATION 3

Making Boxes

What Happens

Session 1: Making Boxes for Cubes Students make open boxes for cubes. Then they decide whether given patterns will make open cube boxes. Finally, they design their own patterns for the boxes.

Session 2: Patterns for Other Solids Students design box patterns, first for rectangular boxes that will hold two cubes, then for triangular pyramids. In both activities, students use visualization skills and creativity in spatial problem-solving contexts.

Mathematical Emphasis

- Exploring 2-D geometric patterns that fold to make 3-D shapes
- Investigating interrelationships between parts of solids
- Improving spatial visualization skills
- Solving problems that require searching for all possible configurations that satisfy given constraints

What to Plan Ahead of Time

Materials

- 1-inch cube: 1 per student (all sessions)
- Construction paper: 1 piece per student (Session 1)
- Scissors: 1 per student (all sessions)
- Tape: 1 roll per pair (all sessions)
- Overhead projector
- Interlocking cubes (for Ten-Minute Math)

Other Preparation

- Duplicate student sheets and teaching resources (located at the end of this unit) as follows:

  Student Sheet 6, Patterns for Cube Boxes: 1 per pair, plus optional transparency

  Student Sheet 7, More Patterns: 1 per pair

  1-inch graph paper (p. 108): 12–15 sheets per pair

  Triangle paper (p. 110): 2 sheets per pair
- Use pattern B on Student Sheet 6 and construction paper to make a sample open box to hold a 1-inch cube. (Session 1)

Session 1

Making Boxes for Cubes

Materials

- 1-inch cubes (1 per student)
- Construction paper (1 piece per student)
- Scissors and tape for every student
- 1-inch graph paper (4–5 sheets per pair)
- Student Sheet 6 (1 per pair, and transparency)
- Overhead projector
- Teacher-made sample box

What Happens

Students make open boxes for cubes. Then they decide whether given patterns will make open cube boxes. Finally, they design their own patterns for the boxes. Their work focuses on:

- determining the shape, number, and spatial relationships between faces of a cube
- finding patterns for open boxes for a cube

Ten-Minute Math: Quick Images As students begin to work with cubes, use the cube images (p. 112) as you continue to do the Quick Images activity. Once or twice during the next few days, outside of math class, use the overhead projector to present a few of these cube images. Students will each need about 10 interlocking cubes to construct the buildings you flash on the overhead; you might place buckets of cubes around the room for easy access.

Activity

Open Boxes for Cubes

Show students a 1-inch cube.

Let's pretend we are box designers working for a box company. For our first project, we must use paper, scissors, and tape to make a box in which this cube fits perfectly. The box should completely cover all but one side of the cube, so we can get the cube in and out easily. No parts of the box should overlap; we are trying to conserve paper.

If students do not understand the task, you might briefly show them an open box that you have made (and kept out of sight) and how the cube fits into it.

Distribute a 1-inch cube and one piece of construction paper to each student to make a box. If some students do not have the manual dexterity to work with inch cubes, you could let them use a larger cube, such as one from the geometric solids set.

Making a box for a cube is not a trivial task for students. They will show a wide range of sophistication in thinking about it.

- Some students wrap the cube as if wrapping a package. They don't see the box as being made up of squares.
- Others trace the square faces of the cube on a piece of paper in a seemingly random fashion, cut out their squares, place them on the cube, and tape them together. Those students see the box as being made up of squares, but do not understand how the squares are connected.
- Other students try to make a pattern of squares that will fold to make the box. If the pattern doesn't work when they try to fold it around the cube, they may start over, or they may cut apart their squares and tape them together in a different arrangement.
- Finally, some students will immediately be able to visualize how the square faces should fit together, and will make one of the standard patterns for a cube.

After all students have completed their boxes, have several individuals show their box and explain how they made it.

Activity

Cube Patterns

Hold up the box that you made. Then disassemble it to show what it looks like flat. Cut it along the four edges that are perpendicular to the bottom so that it looks like this:

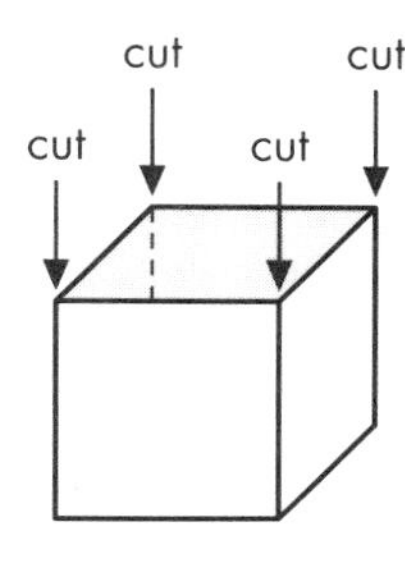

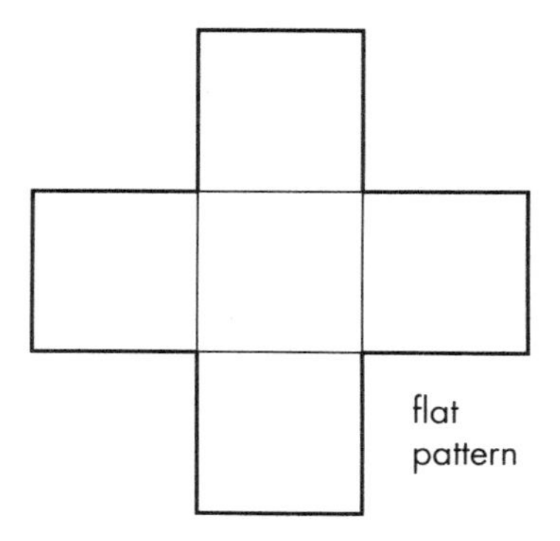

We call this flat piece of paper a pattern for a cube box; it folds up to make a box. What do you notice about this pattern? What are the shapes of its faces? Why is it made from squares?

Students should observe that the sides of the box must match the faces of the cube, and the faces of the cube are squares.

Give each pair of students a copy of Student Sheet 6, Patterns for Cube Boxes.

Which of these patterns do you predict will fold to make an open cube box? Talk about each problem with your partner. Explain why you think a pattern will or won't make an open box.

Write *yes* by the letter of each pattern you think will make an open cube box, and *no* by the letter of each pattern you think will not make a cube box. You can use your cubes to help you predict.

After students have made their predictions, they cut out the patterns to see if they really do make open boxes for a cube. If a pattern does not make an open box, they explain why. When students have finished, have a whole-class discussion about the patterns. You might want to use an overhead transparency of this sheet as a focus for the discussion.

Which patterns did you predict would make an open box for a cube? How did you make your predictions? Which patterns actually made an open box for a cube?

For each pattern that did not work, ask students to explain the reason. See the **Dialogue Box,** Patterns for Cubes (p. 46), for an example of a classroom discussion of this activity.

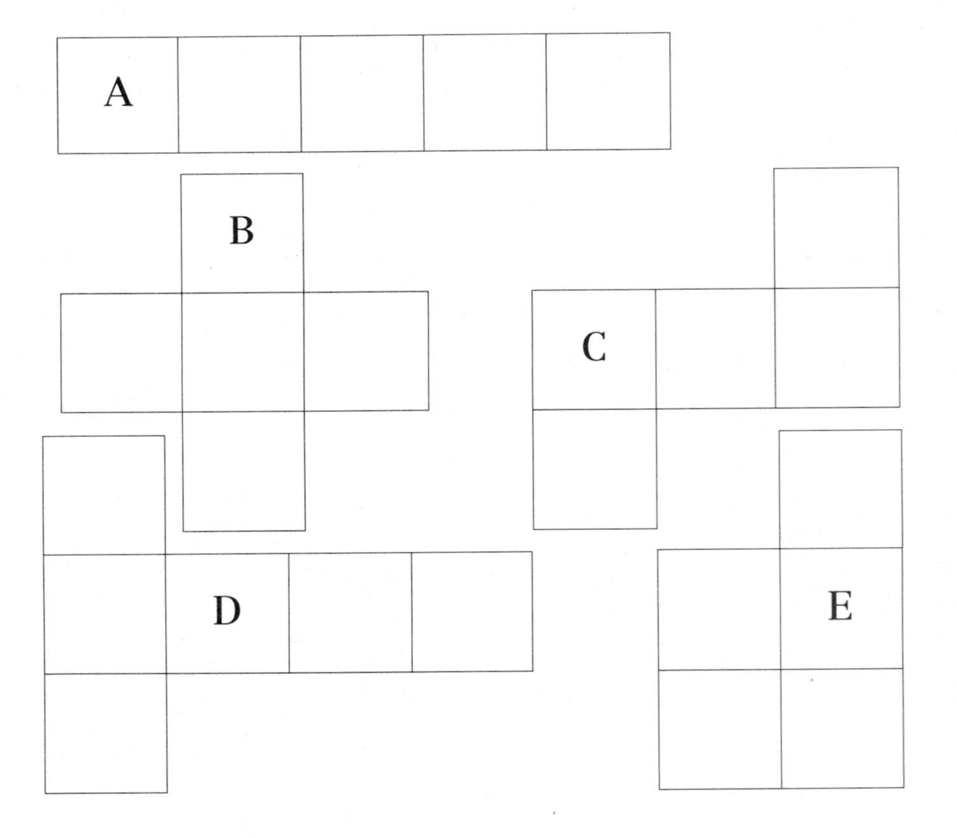

Activity

More Cube Patterns

Distribute 1-inch graph paper to pairs of students.

Our next task as box designers is to find more patterns for open boxes that will fit one cube. Each pattern must follow three rules:

1. It must be made from a single piece of paper.
2. It can be folded only along the edges of the squares.
3. No sides can overlap.
Use graph paper to make your patterns.

As students cut out the patterns, they fold them around cubes to test their ideas, but they don't tape their patterns. Remind them that their job today is to design *patterns,* not boxes.

If a student's pattern fails to make an open cube, ask that student how the pattern could be changed to make it work. Otherwise, many students will not reflect on their errors, but rather will start over repeatedly. Students can cut and paste squares to explore patterns, but when they find one that works, they should draw and cut it out of a single piece of paper.

When the students have found all the patterns they can, have a class discussion. Students might post their patterns on the chalkboard, starting with the ones from the Patterns for Cube Boxes sheet. As students offer new solutions, they can stick those to the board as well.

For each pattern posted, ask the class if they agree that it works. Where there is disagreement, students should justify their beliefs.

Why do you say that this pattern will not work? Will folding it help us check?

Ask students how they devised different patterns.

Duplicate Patterns Sometimes students will post duplicate patterns. In such cases, duplicates should be placed underneath the originals. Discuss with the class which patterns count as duplicates. Most will say that two patterns are the same if one fits onto the other exactly. Some, however, will decide that two patterns are the same if one can be turned, but not flipped, to fit directly on top of the other. It is not really important what criteria students agree on, as long as they discuss the issue and reach a consensus.

Patterns for Cubes

After completing Student Sheet 6, Patterns for Cube Boxes, these students are talking about their answers in a whole-group discussion. Excerpts from their dialogue illustrate some of the ways they try to convince their classmates what the right answers are.

Let's talk about which patterns on the sheet make an open box for a cube.

Yvonne: The answer to the first one is No.

Why do you say that?

Yvonne: All you have to do is fold this up in your mind. There aren't any sides to it.

Do you all agree? [*All agree.*]

Maria: It would fold over [*on itself*] too.

Aaron: It's like a tunnel.

* * * *

What do you think about pattern C?

Seung: I don't think so.

Ricardo: Can I show her [that it does work]? Look. You fold this down. Then you fold it again. Wait, first you fold this. One time. Another time. And then you fold this one. [*Ricardo is not actually folding the paper, but motioning and pointing to the various square faces.*]

* * * *

Pattern D looks interesting. Will it work?

Chantelle: No. Not if you want to make an open box.

Cesar: Yeah, it has a top.

Jeremy: It has 6 squares; it should have 5.

* * * *

What about pattern E?

Kate: No way. If you put it [the cube] here, anywhere you put it—it would not work. Like there. You fold this up and then you fold this up, what would you do with this square? This one? This one?

Note that some students "see" the folded box without the cube present, while others need to use a cube to visualize the results of folding the patterns.

Session 2

Patterns for Other Solids

What Happens

Students find box patterns, first for rectangular boxes that will hold two cubes, then for triangular pyramids. In both activities, students use visualization skills and creativity in spatial problem-solving contexts. Their work focuses on:

- determining the shape, number, and spatial relationships between faces of solids
- finding multiple solutions to problems

Materials

- 1-inch cubes (2 per pair)
- Scissors and tape for every student
- Student Sheet 7 (1 per pair)
- 1-inch graph paper (8–10 sheets per pair)
- Triangle paper (2 sheets per pair)

Activity

Teacher Checkpoint

Patterns for Two-Cube Boxes

Distribute Student Sheet 7, More Patterns, along with inch cubes, 1-inch graph paper, and triangle paper to each pair of students.

For the first problem, students (working in pairs) use the graph paper to make patterns for open boxes that hold two 1-inch cubes. They may tape two inch-cubes together to help them design and check their patterns. As before, each pattern must satisfy the following rules:

1. It must be made from a single piece of paper.
2. It can be folded only along the edges of the squares.
3. No sides can overlap.

Students should not tape their patterns into boxes at this point—they are designing patterns, not boxes. However, they should cut out each pattern, fold it, and place the inch cubes inside to be sure that it works.

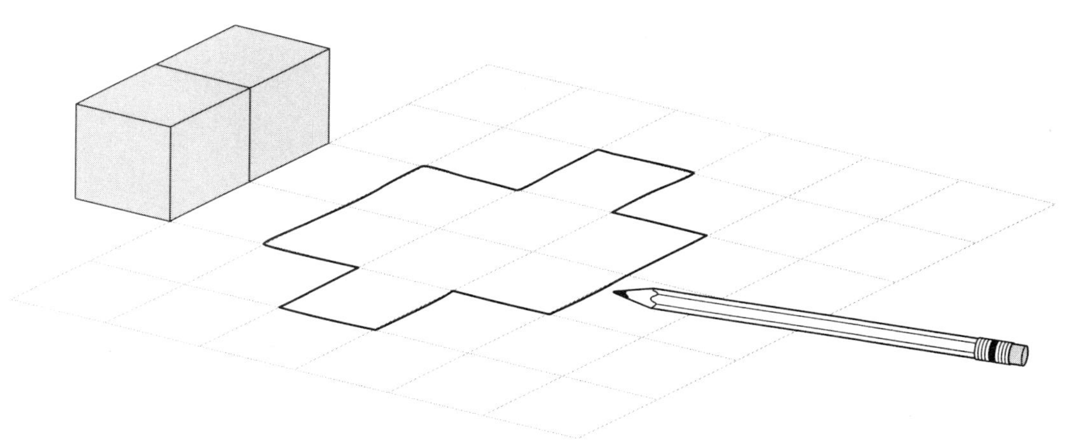

See the **Dialogue Box,** Making Patterns for the Two-Cube Solid (p. 51), for a description of how one teacher encouraged students to keep searching for patterns, and for examples of patterns students have designed.

When students seem to have found all the patterns they can, have a class discussion similar to that for the one-cube boxes. Students show their patterns and explain why they work. Students who disagree can challenge any pattern. Of course, disagreements can be settled by folding the patterns to see if they make an appropriate box.

How many squares must be in a pattern? Why?

Students should discover that there can be either 8 or 9 squares, depending on whether the open top is 1 or 2 squares across.

Examining Student Solutions Circulate around the room examining students' attempts at solutions. Did they follow the rules for patterns? Did they make several patterns? What strategies did they use? Did they use cubes, or could they visualize the patterns without the cubes? Can they alter a pattern that does not work to make one that does? Can they alter a pattern that works to make another that works? You will find that students use a wide variety of strategies. As long as they can find several patterns that work, they are doing fine.

Because students are so creative at designing patterns, it's nice to find ways to display their work. Some teachers set up a section of a bulletin board or chalkboard where students can attach their patterns. One student made her own book of patterns. She glued only the bottom squares of the patterns to the pages of her book so the patterns could be folded into boxes by readers.

Activity

Patterns for Triangular Pyramids

Students are next asked to use triangular grid paper to find patterns for *closed* boxes that will hold a triangular pyramid (all four of the faces are equilateral triangles). They are allowed to make different-sized pyramids. They should fold, but not tape, their patterns into boxes to check their work.

As when making two-cube box patterns, the students work in pairs. When they have a variety of patterns, post their different ideas on the board. Hold a whole-class discussion of their work, testing any patterns that are challenged, and encouraging the makers to explain and defend their patterns.

Several student-designed patterns are shown below. The patterns make two different sizes of pyramid, but each has four faces that are equilateral triangles.

Session 2 Follow-Up

Homework

Challenge students to try to make more patterns that will make a two-cube box.

Extension

Patterns for Other Figures Some students enjoy trying to make patterns that will fold to make other geometric solids. They may want to use Student Sheet 5 (which pictures 12 geometric solids) for reference, and might need to use wooden solids to trace the faces.

pattern for triangular prism

Making Patterns for the Two-Cube Solid

While working together to create patterns that will make an open box to fit a two-cube solid, Tyrell and Yoshi have made one pattern:

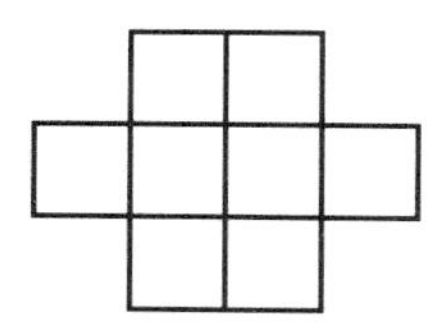

Now their teacher is encouraging them to find more patterns for the same shape.

Good. How did you think of that pattern?

Tyrell: It's like the cross for the cube.

Oh, that was good thinking. Do you think there are any more patterns for two cubes?

[*Both boys hesitate.*]

There were quite a few patterns for a single cube.

Tyrell: But that was different. There are two cubes now. [*He tries drawing a new pattern.*]

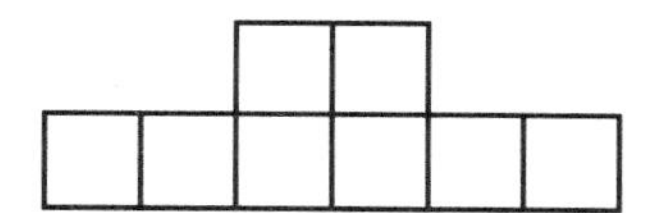

Yoshi: Yeah, that works. And you can have another pattern if you just move these over. [*He indicates a movement of the top two squares from positions 3 and 4 to positions 4 and 5 in the row of six squares. Yoshi then draws and cuts out the following pattern:*]

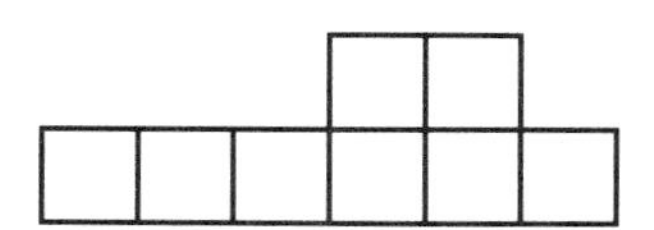

It's the same as putting them here. [*Yoshi indicates putting the top two squares above positions 2 and 3 in the row.*]

Tyrell: Here's another one.

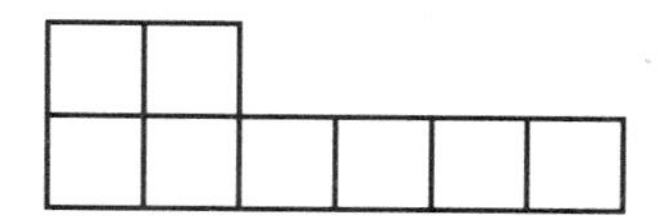

Do you think there are any others?

Yoshi and Tyrell: I don't think so. [*The boys seem to be focused on their "move the two squares" strategy and cannot readily see other possibilities.*]

Keep looking; see if you can find more.

Some other patterns that students have designed for open two-cube boxes are shown below. The shaded squares show the bottom of the box. Note that the bottom (and thus the open top) can be either one or two squares.

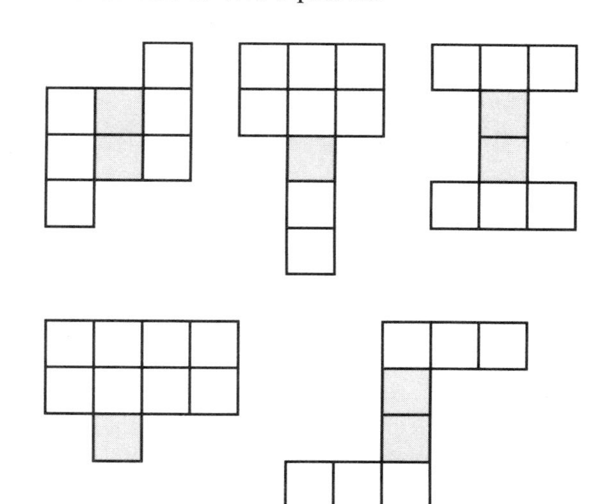

INVESTIGATION 4

How Many Cubes in a Box?

What Happens

Session 1: Finding the Number of Cubes in a Box Students examine patterns for open boxes drawn on square grids. They predict the number of cubes the boxes will hold; then they check their predictions by building boxes from the patterns and filling them with cubes.

Session 2: Twelve-Cube Boxes Students use square grids to design open rectangular boxes that hold exactly 12 cubes. This activity causes students to reflect more deeply on the structure of box patterns.

Session 3: Patterns from the Bottom Up Students complete designs for box patterns, given the shape of the bottoms of the boxes.

Mathematical Emphasis

- Understanding how a pattern for a rectangular box folds to make the box
- Predicting the number of cubes that fit in a box by examining a pattern that makes the box
- Determining the number of cubes that fit in a rectangular box
- Understanding the structure of rectangular prism arrays of cubes

What to Plan Ahead of Time

Materials

- Scissors: 1 per student (Sessions 1–3)
- Tape: 1 roll per pair (Sessions 1–3)
- Interlocking cubes: 60 per pair, stored in resealable plastic bags or plastic tubs with lids (Sessions 1–3) (See note related to cube size, below.)
- Overhead projector

Note on Cube and Grid Size

If you are not using the ¾-inch cubes supplied with the grade 3 materials kit, be sure to use graph paper that matches the size of your cubes. Also, you will need to adjust the following pages, which are based on a ¾-inch grid:

- Student Sheet 8, How Many Cubes? (4 pages)
- Student Sheet 9, Making Boxes from the Bottom Up (3 pages)
- Student Sheet 10, An 18-Cube Box Pattern
- Demonstration Box Pattern (p. 106)
- Discussion Box Pattern (p. 107)

For example, if you use 2-cm cubes, you will need to enlarge these sheets by about 5% on a copier. Experiment with one enlarged copy first, and test it carefully to be sure the squares match the faces of the cubes students will be using.

Other Preparation

- Sort 60 interlocking cubes into plastic bags or tubs for each pair. (Session 1)
- Duplicate student sheets and teaching resources (located at the end of this unit) as follows:

For Session 1

Student Sheet 8, How Many Cubes? (4 pages): 1 per pair

Demonstration Box Pattern (p. 106): 1 copy (cut out and fold, but do not tape it, to make an open box)

Discussion Box Pattern (p. 107): 1 paper copy and 1 transparency

For Session 2

¾-inch graph paper (p. 109): at least 2–3 sheets per student, more for homework

For Session 3

Student Sheet 9, Making Boxes from the Bottom Up (3 pages): 1 per student, and one set of overhead transparencies

Student Sheet 10, An 18-Cube Box Pattern: 1 per student

Session 1

Finding the Number of Cubes in a Box

Materials

- Student Sheet 8 (1 per pair)
- Interlocking cubes (24 for teacher demonstration; 60 per pair in bags or tubs)
- Demonstration Box Pattern (cut out)
- Transparency of the Discussion Box Pattern, plus 1 paper copy
- Overhead projector
- Scissors and tape for every student

What Happens

Students examine patterns for open boxes drawn on square grids. They predict the number of cubes the boxes will hold; then they check their predictions by building boxes from the patterns and filling them with cubes. Their work focuses on:

- finding methods to predict the number of cubes that will fit in the box made by a given pattern on a square grid
- finding methods to determine the number of cubes that fit in a box constructed from graph paper

Activity

Predicting the Number of Cubes in a Box

All kinds of things are packaged in boxes—toys, cereal, games. We're going to design some boxes. First, we are going to study how much stuff will fit in the boxes we make. We will look at boxes without tops because they are easier to understand.

Show the Demonstration Box Pattern and fold it to make a box without a top, but don't tape it. Note that the lines should show on the outside.

Here is a pattern for a box. When I fold it up like this, it forms a box without a top. How many of these cubes do you think will fit inside the box? [*Show an interlocking cube.*]

Tell me how you made your predictions.

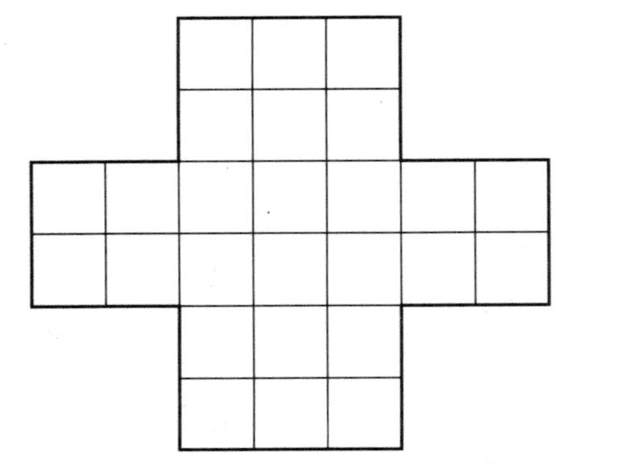

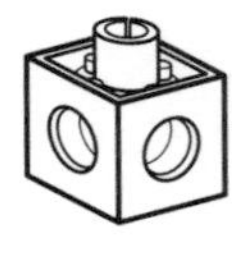

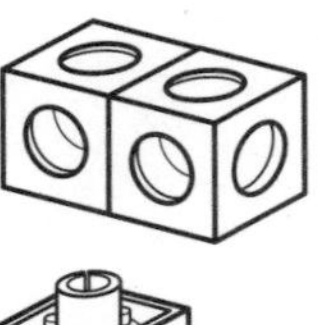

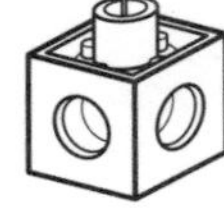

Students' strategies will vary. For the Demonstration Box Pattern, some will predict 12 cubes, saying there are 6 cubes on the bottom and 6 more on the top layer. Others might count the squares in the pattern, getting 26. It may seem that they have misconstrued the problem, thinking about squares rather than cubes. However, many are actually thinking that a cube fits on each square and that when the pattern is folded, these cubes will fill the box exactly.

After all predictions and strategies for making predictions have been discussed, check the predictions by folding and taping the pattern into a box and putting the cubes inside. (Note that the faces of ¾-inch interlocking cubes match the size of the squares in the Demonstration Box Pattern and the patterns on the student sheets, so the cubes should fit into the boxes perfectly *if they are connected.*)

Why was your prediction different from the number of cubes that actually fit in the box?

Activity

Making and Checking Predictions

Distribute the four pages of Student Sheet 8, How Many Cubes? to pairs of students who will be working together. Each page of this handout contains a pattern for a different rectangular box, open at the top. Students are to work with one pattern at a time.

Starting with pattern A, students first predict how many cubes will fit in the completed box, writing their predictions at the center of the pattern. Be sure students understand that their predictions should not be mere guesses, but should be based on their analysis of how the cubes will fit into the folded box.

Students then cut out and fold pattern A into a box and fill it with interlocking cubes to determine how many fit inside. Be sure students are connecting the cubes—otherwise they will not fit into the box properly.

Students then go on to the next pattern. They make a prediction for the next box, then build it, then check their prediction; they do the same for the third box, and finally the fourth. This procedure encourages them to keep refining their prediction strategies.

Student Strategies As you circulate around the room, investigate the strategies students use to determine the actual number of cubes that fit in the boxes. You will find students making errors not only when they make predictions but also when they use the cubes to fill the boxes. See the **Teacher Note,** Strategies for Finding the Number of Cubes in a Box (p. 57), for ideas about how students approach this problem.

Activity

Discussing Student Strategies

The final activity in this session enables you to more closely observe the strategies students are using to predict the number of cubes that will fit in a box made from a pattern. This should help you better understand how individual students are making sense of the problems.

Show the Discussion Box Pattern transparency, and ask each student to predict how many cubes will fit in the box that it makes. Students write their predictions and a description of how they made the prediction on a sheet of paper and hand it in. The **Teacher Note,** Strategies for Finding the Number of Cubes in a Box (pp. 57–58), includes some examples of students' written descriptions.

❖ **Tip for the Linguistically Diverse Classroom** Give each student who may have difficulty describing his or her prediction a copy of the Discussion Box Pattern. Students can mark on this sheet to show how they made their predictions. For example, two students might show their answers as follows:

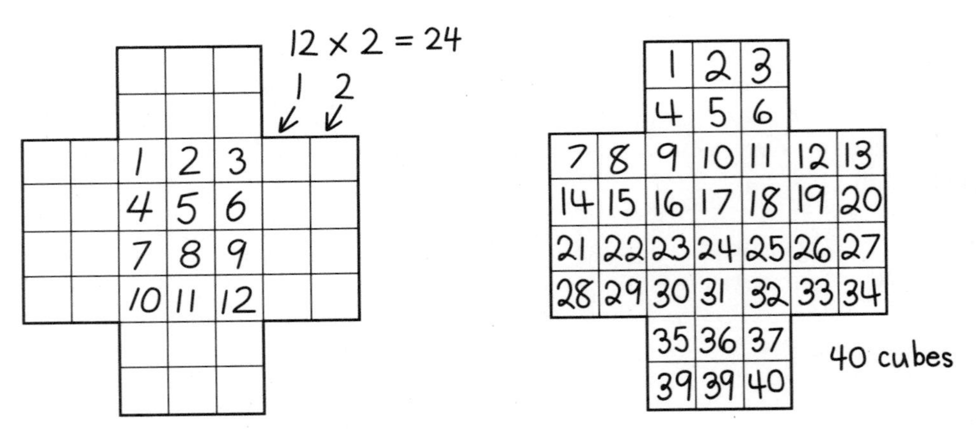

You can easily tell from their markings that one student understands the problem and the other one still does not.

Follow up with a whole-class discussion, asking students to give their predictions and describe their strategies for making them. After all predictions have been discussed, fold a copy of the pattern and fill it with cubes to test the predictions. Ask students how to determine the number of cubes that actually fit in the box. Try the different methods suggested and have the students discuss any discrepancies.

Strategies for Finding the Number of Cubes in a Box

Teacher Note

The strategies students use to determine the number of cubes that fit in a completed box will vary greatly. Some students will count the cubes that fit into the box one by one. Some will count the cubes in a layer (not always a horizontal layer), then use repeated addition or skip counting to find the total. Others will use multiplication to determine how many cubes are in a layer, then use multiplication again to find the total number of cubes.

Some students, however, will count squares. They might count squares in the pattern. Or they might fill a box with cubes, then take the set of connected cubes out of the box and count the total number of squares visible on all sides. It is essential that you help these students correctly determine the number of cubes.

The most basic reliable method for determining the number of cubes is to count them one by one as they are taken apart. However, before students resort to this method—which they can all employ—you might model a more efficient method. For instance, you could separate the rectangular cube array into layers, asking students to count the cubes in a layer, then successively stack layers as students keep track of the total.

How many cubes do you think are in the bottom layer? How do you know? How many cubes are there after I add this layer? What about after I add this layer? So how many cubes are in the whole box?

We are presenting this task to students as a problem to solve. It is important that they develop their own workable strategies to solve it, and that they gain the experience they need to make sense of rectangular cube configurations. Thus, it is essential not to impose on students any particular approaches to predicting or determining numbers of cubes. If you do, most students will learn the suggested approaches by rote.

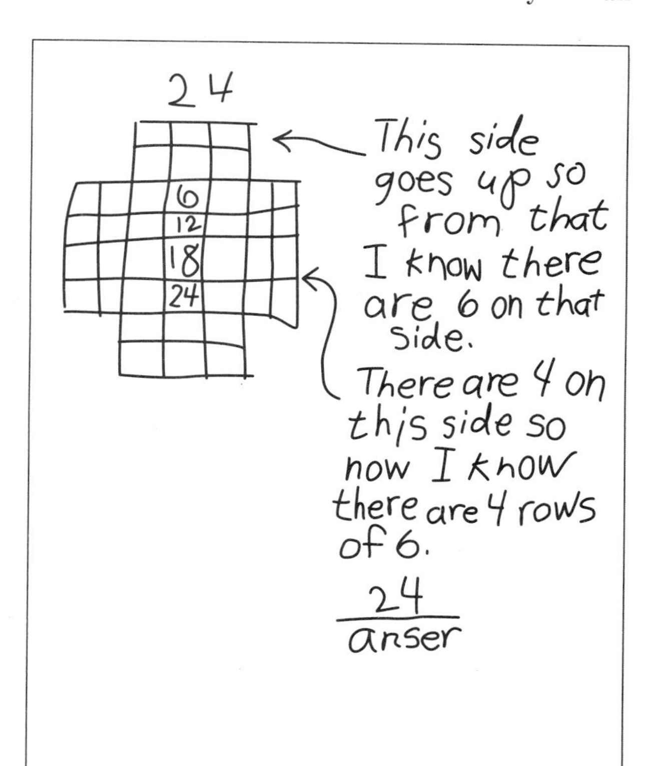

I think that the answer is 24. The reason I think that is because the sides are two layers and the middle is twelve.
Two × 12 = 24.

Continued on next page

Teacher Note

continued

Instead, allow class discussions and your modeling to inform students of more efficient strategies than the ones they are currently using. For instance, after a discussion of strategies, if the idea of using layers has not been mentioned, model it as described above. However, don't expect all students to adopt this strategy. They should, and will, adopt more efficient strategies when they feel comfortable with them.

The precise method students employ is not so important; that they employ a method that makes sense to them and gives the correct number of cubes is what is essential.

There is, however, one strategy that we distinctly do not want to encourage: the procedure of finding the number of cubes in a 2 by 3 by 4 box by writing the three-factor product $2 \times 3 \times 4$ and multiplying. This is essentially using the formula, *length* × *width* × *height* = *volume*. We have found that most students in elementary school simply memorize the formula—they have no idea why it works. In fact, finding the number of cubes in a box by finding the number of cubes in a layer and multiplying by the number of layers is a much more powerful and general way of thinking about this problem.

I counted the numbers in the middle and it was 12. Then I looked on the side and I counted up the sections and it was two. then what I did was multiply them togather and it was 24.

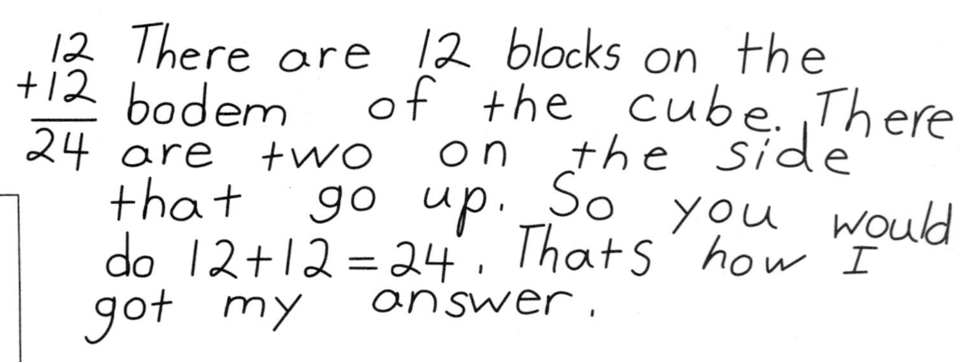

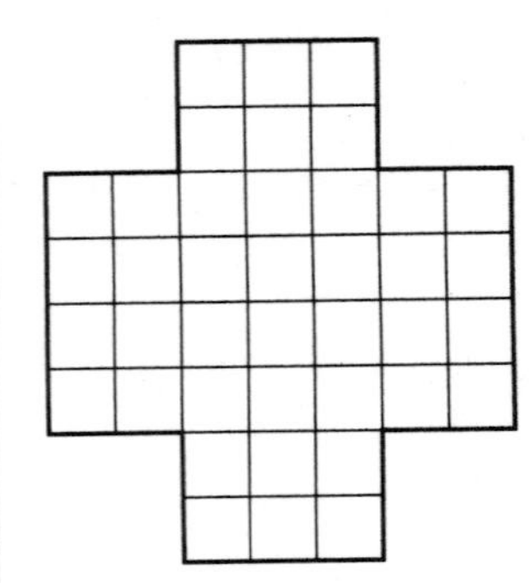

Session 2

Twelve-Cube Boxes

What Happens

Students use square grids to design open rectangular boxes that hold exactly 12 cubes. This activity causes students to reflect more deeply on the structure of box patterns. Their work focuses on:

- finding methods to predict the number of cubes that will fit in the box made by a given pattern
- designing patterns for boxes that will hold a given number of cubes

Materials

- ¾-inch graph paper (at least 2–3 sheets per student, another 2–3 for homework)
- Interlocking cubes
- Scissors and tape for every student

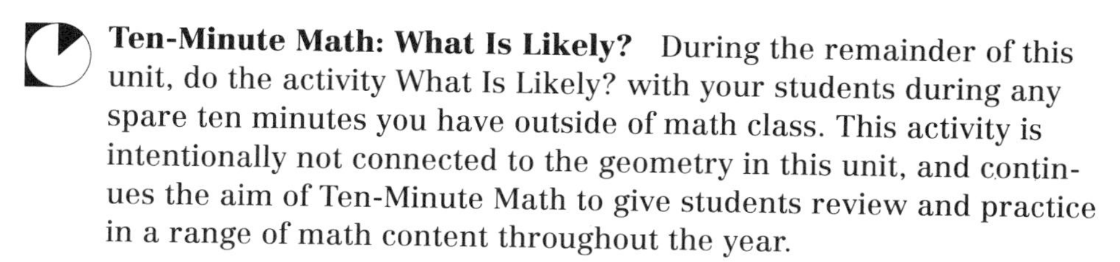

Ten-Minute Math: What Is Likely? During the remainder of this unit, do the activity What Is Likely? with your students during any spare ten minutes you have outside of math class. This activity is intentionally not connected to the geometry in this unit, and continues the aim of Ten-Minute Math to give students review and practice in a range of math content throughout the year.

Try to do this activity three or four times as you finish this unit. You will need a clear container filled with two colors of blocks, beans, tiles, or some other material of similar size and shape. The first one or two times you do the activity, put in the container much more of one color than the other, for example, 18 red cubes and 2 yellow ones.

Start by showing the bowl or jar to the students. Ask them to predict which color they will get more of if they draw ten objects out of the container (with their eyes closed). Students then draw ten objects from the bowl, replacing the object after each draw. Record the color of each object drawn.

Discuss what happened. Then invite students to try it again and discuss the results. For full directions and variations, see p. 79.

Activity

Designing Boxes to Hold 12 Cubes

Yesterday you had four box patterns, and you figured out how many cubes would fit in those boxes. Today you're going to design your own patterns for boxes.

The problem is to design open rectangular boxes that will hold exactly 12 cubes. The goal is to make as many different boxes as possible. You can use cubes to help you make your designs.

There are many different boxes that students can make to solve this problem. We consider two boxes that hold the same cube configuration different if their open tops are different. For instance, there are two different boxes that will contain a 1-by-1-by-12 cube configuration. One has a 1-by-1 open top; the other has a 1-by-12 open top.

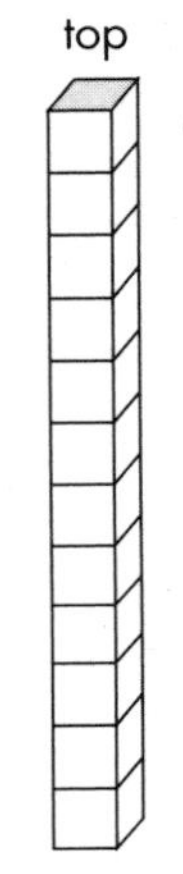

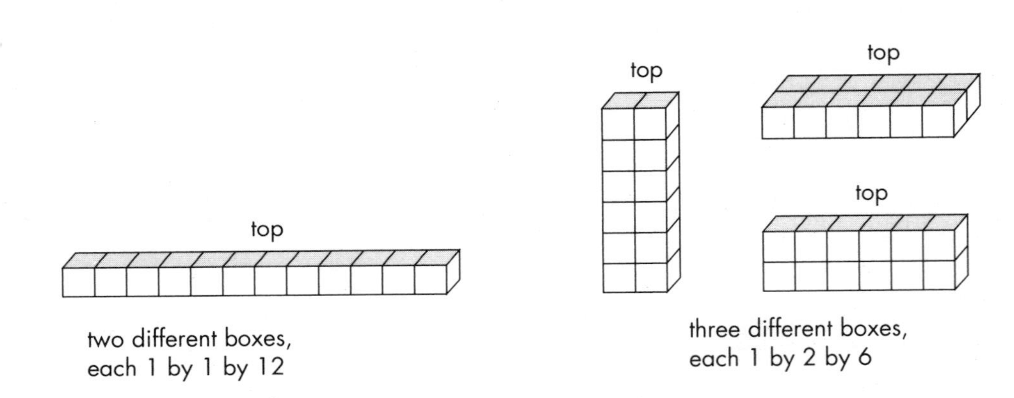

two different boxes, each 1 by 1 by 12

three different boxes, each 1 by 2 by 6

Similarly, there are three boxes that contain a 1-by-2-by-6 cube configuration—one with a 2-by-6 bottom, one with a 1-by-2 bottom, and one with a 1-by-6 bottom.

Students work in pairs on this activity, but all should have their own supply of graph paper with which to search for patterns. To find 12-cube boxes, students should draw their ideas for patterns on ¾-inch graph paper, cut them out and fold them, then fill them with cubes to test for fit. Patterns that work should be taped together into boxes and saved.

Students should be encouraged to use cubes whenever they need them, from the early design stage to the checking of final patterns. As students are taping their boxes, you might suggest that they leave one face untaped, making it easier to put in and take out the cubes.

Encourage students to make their boxes out of whole pieces of paper rather than separate pieces. At the same time, explain that some boxes for

12 cubes will require that they tape several pieces of paper together. For example, neither of the boxes for a 1-by-12 cube configuration can be made from a 10-by-12 grid without cutting and taping.

Even though students worked with patterns in the previous activity, figuring out how to make the patterns themselves is difficult for many. Therefore, student strategies will vary greatly in sophistication. See the **Dialogue Box,** Twelve-Cube Box Strategies (p. 62).

For students who are having difficulty, you might suggest that they build a rectangular building from 12 cubes first, place it on the grid, then try to draw the pattern for the box. Give students plenty of time to devise multiple solutions.

Occasionally, students will design nonrectangular boxes. You might praise their creativity, but explain that rectangular boxes are needed for shipping and stacking.

When a student's pattern does not work, encourage him or her to not simply discard it, but to think about how to change it to make it work.

How many cubes does your pattern hold? Could you change it so that it would hold 12 cubes? How?

Sometimes, mentioning layers will help students design the box patterns.

How many cubes will there be in the bottom layer? How many layers will there be?

It is clearly not necessary that students discover all possible boxes. What is important is that they remain productively engaged in searching for new boxes, that is, that they are generating and evaluating new ideas for boxes, and that they can decide when two boxes are the same or different.

Session 2 Follow-Up

For any given 12-cube box, several different patterns can be used to make that box. As a homework project that can last several days, you might ask students to find different patterns for the same 12-cube boxes they have found in class. Be sure to send home sufficient amounts of graph paper for this activity.

Post examples of the boxes on a bulletin board. Below each box, students post the different patterns they find that will make that box.

Twelve-Cube Box Strategies

The following discussion among students working to make their 12-cube box patterns demonstrates the range of strategies you are likely to encounter in your class.

What methods are you using to make your boxes?

Tamara: I just drew a rectangle that had 12 squares, then I put on the sides.

Dominic: I made a building with 12 cubes in it. Then I put it on my paper and traced the bottom. I looked at the sides of the building and could tell how to draw the sides of my box.

Elena: You count the cubes in the bottom and it has to be a factor of 12. Then for the sides, you have to do whatever [number] is times that factor that makes 12. Say there were 6 in the bottom, 6×2 is 12, so you have to make each side go up two squares.

Kate: I just sort of look at it. I draw a pattern, then fold it up to see if it works. If it doesn't work, I try again.

Consider how these different students are approaching the problem. Tamara's strategy is common. The drawback with this strategy is that it will not produce multilayer boxes. In fact, some students who use this strategy always ignore the height of the sides, discovering their error only after they attempt to fill their box with cubes.

Dominic's strategy is especially useful for students having difficulty, and could be suggested to them as a good approach.

Elena's strategy is quite sophisticated—she seems to be using information she learned in the multiplication and division unit (including the term *factor*). In fact, it is not uncommon for students to comment that they can solve this problem "because I know my multiplication pairs."

Finally, Kate has some difficulty expressing her strategy accurately. It might seem as if she is using random trial and error. However, because she was successful in making several boxes, her trial and error seems to have been guided by some intuitive ideas that she either could not express or wasn't explicitly aware of.

Students can devise some amazingly creative patterns, especially when prompted to find more than one pattern for the same 12-cube box, as suggested for homework (p. 61). For example, the diagram below shows two different patterns students made for the 1-by-1-by-12 box.

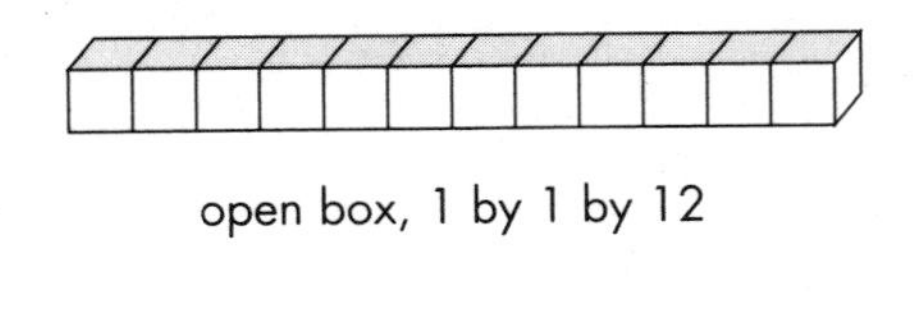

open box, 1 by 1 by 12

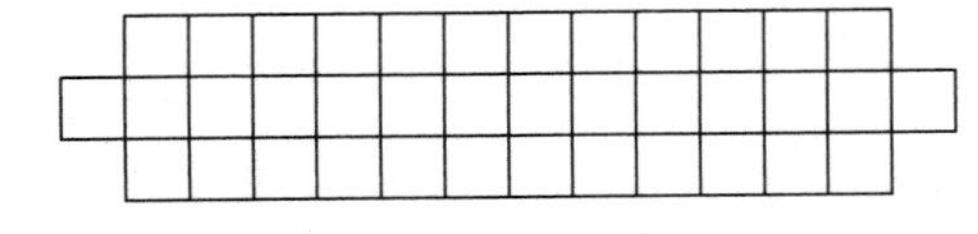

Jennifer's pattern

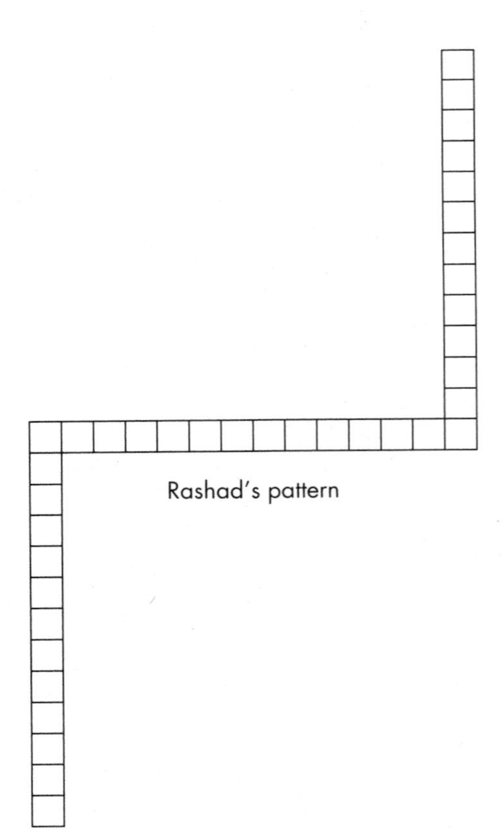

Rashad's pattern

Patterns from the Bottom Up

What Happens

Students complete designs for box patterns, given the shape of the bottoms of the boxes. Their work focuses on:

- designing patterns for boxes that will hold a given number of cubes
- understanding the relationship between the configuration of cubes that fit in a box and the location of squares that appear on the sides and bottom of the box

Materials

- Transparencies of Student Sheet 9 (3 pages)
- Student Sheet 9 (1 per student)
- Student Sheet 10 (1 per student)
- Scissors and tape for every student
- Interlocking cubes

Activity

From Bottom to Sides

As you may have discovered in the preceding session, designing box patterns is a difficult task. The goal of this session is for students to gain more experience designing such patterns and to develop strategies for designing patterns through sharing ideas and through teacher guidance.

Give each student the three pages of Student Sheet 9, Making Boxes from the Bottom Up. Show the first pattern on the overhead projector as you explain the activity.

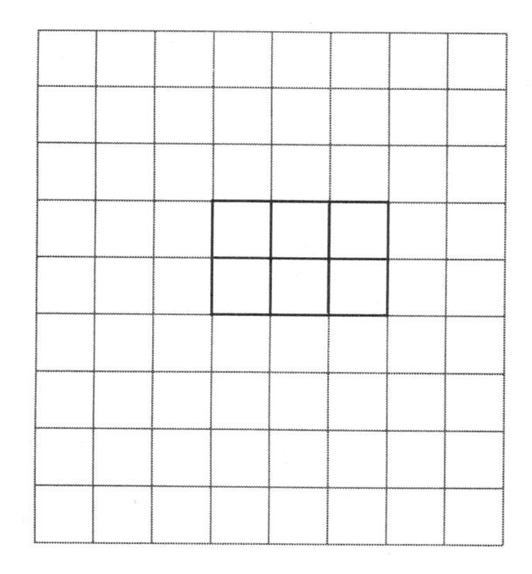

This is the bottom of a rectangular box that contains exactly 6 cubes. The box has no top. Draw the sides to complete the pattern for the box. You can use cubes to help you if you like.

After students have spent some time working, ask them to share their strategies.

Somebody show us on the overhead how you completed the pattern. How did you decide what the sides would look like? How many layers does your pattern have? How did you know how many layers there would be? Did anybody come up with a different pattern?

Even though there is only one pattern that works, let students show and discuss whatever patterns they have designed.

Now each of you make a box from your pattern and test it by filling it with cubes. Was your prediction for the pattern correct?

If the pattern predicted on the overhead was not correct, ask a student to draw the correct pattern on the overhead.

Repeat the above procedure for each problem. While students are working individually, you will have a chance to help those who are having particular difficulty.

To make or even to understand the patterns, some students will need to make the cube configurations that fit in the boxes first. For instance, for the first problem, they would make a 2-by-3 rectangle of six cubes, place it on the 2-by-3 pattern for the bottom of the box, and try to figure out what the sides would look like. If they don't see how to proceed, you might point to one of the lateral sides of the building and ask:

How tall is the building, how many layers? How many cubes do you see on this side? Can you draw this side on the paper?

Repeat these questions for the other sides as necessary.

Activity

Assessment

An 18-Cube Box Pattern

Distribute Student Sheet 10, An 18-Cube Box Pattern. The problem on this sheet is just like those on Student Sheet 9, except that instead of cutting out and taping their pattern to check, students write what strategy they used in making their pattern. For example, did they simply see in their minds what the pattern looked like? Did they make a cube building first and place it on the paper to figure out the sides?

❖ **Tip for the Linguistically Diverse Classroom** Be sure students have the option of showing and explaining to you how they completed the pattern, rather than writing about it. Students can use drawings and cubes to demonstrate their thinking. For recording purposes, you can help them write their approach at the bottom of their sheet.

Collect this sheet to review students' work. As you assess their work, note whether students drew the patterns correctly, and the method they used to make the pattern.

Students who can make the correct pattern, no matter what method they use, are doing well. Students who must use cubes to make their patterns should be permitted to do so as long as they feel it necessary. Still, you can enhance the development of their visualization skills by encouraging them to first predict the patterns without using cubes, then to check their answers with cubes.

For students who are unable to make correct patterns, you can make up additional problems on graph paper that matches the size of your cubes. For instance, you could draw a 4-by-5 array and ask students to complete the pattern so that the resulting box will hold 20 cubes. Similarly, you could draw a 5-by-1 array and ask students to complete the pattern to make a 15-cube box.

It is not necessary that students master these problems before moving on to the project in Investigation 5. In fact, that project will provide a context in which many students will become more motivated to develop a good technique for making patterns. Nevertheless, this assessment will help you identify students who will need extra guidance as they make box patterns for their project.

Session 3 Follow-Up

Homework

As a homework assignment, ask students to make patterns for rectangular open boxes that contain exactly 20 cubes. Be sure to send home sufficient amounts of graph paper. To conserve paper, you could use paper with smaller squares, thereby getting more patterns out of each sheet; however, keep in mind that students will be unable to check these patterns using their larger cubes.

Building a City

What Happens

Sessions 1, 2, 3, and 4: Making a Box City Students build a model of a city—a set of open-box buildings made from patterns that they draw on graph paper. In addition to the model, students submit a set of patterns for their city buildings, as well as a written description of the rationale that guided their choice of buildings.

Mathematical Emphasis

- Problem solving involving planning, trial-and-error processing, analysis, and visualization
- Predicting the number of cubes that will fit in the box made by a given pattern
- Designing patterns for boxes that will hold a given number of cubes
- Using appropriate computation techniques to determine the total number of cubes that will fit in different boxes

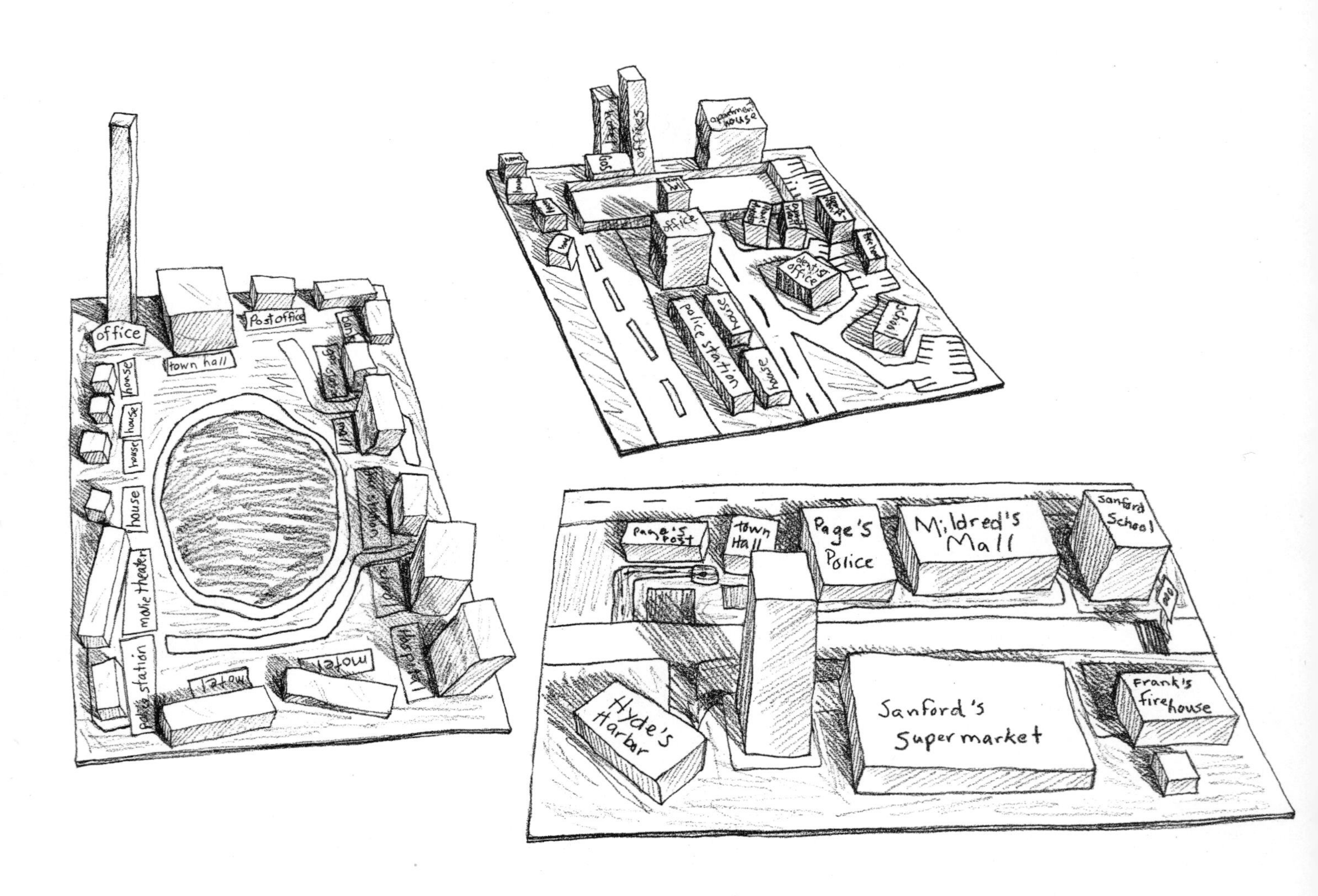

What to Plan Ahead of Time

Materials

- Interlocking cubes: 60 per pair in bags or tubs
- Scissors: 1 per student
- Tape: 1 roll per pair
- Calculators: 1 per pair
- Quart-size resealable plastic bags for storage: 1 per student (optional)

Other Preparation

- For this project students will need a large supply of 3/4-inch graph paper. You can duplicate the sheet on p. 109, making at least 18 copies per student pair. However, because a larger size (11-by-17-inch) works better for this activity, consider taping together two of the smaller sheets (cutting and taping carefully along the edge of one grid) to make a grid with 12 by 20 squares and duplicating this on large paper, making at least 9 copies per student pair.

  If you want to use commercial graph paper, be sure you find a grid that matches the size of your cubes.
- Make a wall chart of the City Building Code rules listed on p. 69, modifying rule 4 to "two sheets" if you are supplying students with 11-by-17-inch graph paper.

Sessions 1, 2, 3, and 4

Making a Box City

Materials

- Interlocking cubes
- Large supply of $\frac{3}{4}$-inch graph paper for every student
- Scissors and tape for every student
- Quart-size resealable plastic bags for storage (1 per student, optional)
- Calculators (1 per pair)
- Chart of building code

What Happens

Students build a city of open-box buildings made from patterns that they draw on graph paper. They submit a set of patterns for their city buildings as well as a written description of their plan for the city. Their work focuses on:

- designing patterns for boxes that will hold a given number of cubes
- determining the number of cubes that will fit in the box made from a given pattern

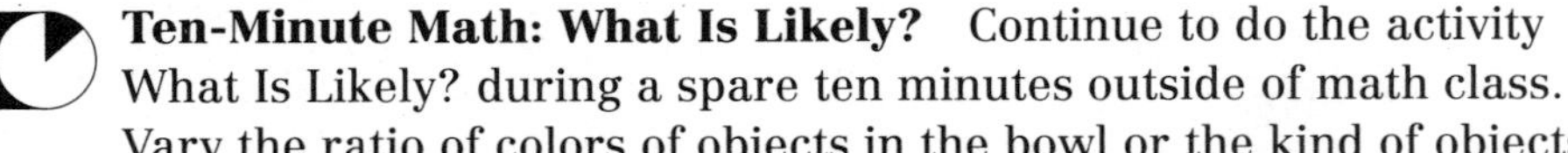

Ten-Minute Math: What Is Likely? Continue to do the activity What Is Likely? during a spare ten minutes outside of math class. Vary the ratio of colors of objects in the bowl or the kind of objects you are using. For full directions and variations, see pp. 79–80.

Activity

Introducing the Problem

Post the City Building Code chart where everyone can see it.

Today we are starting a project that will last three or four days. You are going to work as architects, planning a new city. You will make models of your city buildings from boxes.

The resources for building the model city are limited; you are allowed to use only four sheets of graph paper [or two sheets of the larger size] to make all of its buildings.

Also, the people who will live in the city want to know how much room or space will be in the city. So you must figure out how many cubes each of your buildings holds, and then how many cubes your whole city will hold.

As you design your city, be sure to follow the ***building code,*** **which means the rules for building in this city. They are posted on this chart.**

CITY BUILDING CODE

1. Each building must have the shape of a rectangular prism.
2. Each building must be made from a box with no bottom.
3. Each building must be made from a single sheet of paper.
4. Only four sheets of graph paper can be used to make all the buildings in the city. You may get more than one building from a single sheet.
5. You must have at least three differently shaped buildings.

Read through the building code with the students.

❖ **Tip for the Linguistically Diverse Classroom** Be sure all students understand the building code that governs this assignment. For example, demonstrate with geometric solid figures that only rectangular prism shapes can be used for buildings. Show graph paper to demonstrate that each building must be made from a single sheet of paper, and that all their buildings must be made from a total of four sheets.

Explain that, as architects, students must submit three things for their cities:

- Sheets of graph paper that show all the patterns they used to make the buildings, how many cubes fit in each building, and the total number of cubes that fit in the whole city—this is the plan for the city.
- A set of buildings made from these patterns, taped to a large piece of construction paper or cardboard—this is the model for the city.
- An explanation of why students designed their city as they did.

You might want to list these requirements on the board as a reminder throughout the investigation, or simply review them with the students as they near completion of the project.

Students will at first be working individually as they design buildings that can be made from their four sheets and determine how much room is in each building. Later, they will be working in pairs, combining their work as they decide on a set of plans for a model city.

Explore with students some of the things they may want to think about as they plan their cities.

There are many different ideas that can guide your design. For example:

- **Some designers want to make a city that will contain as much room or space inside it as possible. They want the city buildings to hold the most cubes possible.**
- **Some designers want their city to contain lots of separate buildings, so the people living in it don't feel crowded.**
- **Some designers want tall buildings with small bottoms because land in their city is very expensive.**
- **Some designers want the new city to look like the city they live in.**
- **Some designers want their city to be fun to visit, for both children and adults.**

What are some other things you might think about in planning your city?

Activity

Planning and Building the City

Give each student four sheets of letter-size graph paper or two sheets of the larger (11-by-17-inch) paper. Remind students that the pattern for each building must be made from a single piece of paper, cut from these sheets. Leftover scraps can be used to make more patterns, as long as they are not taped together.

Initial Plans Students begin working individually on the project, using their graph paper to draw a set of building patterns. When the students have drawn all their patterns, they should cut them out and determine how many cubes each holds. Interlocking cubes should be available for students to use in making patterns and checking numbers of cubes. Students should also find the total number of cubes for all their buildings; encourage the use of calculators.

At the end of each session, have the students save their patterns and calculations for the next day. Large resealable plastic bags are useful for this purpose.

As you circulate around the room, be sure that students are following the Building Code. Ask students to tell you how they determined the number of cubes that fit in their buildings. Help them recognize any mistakes they have made. If some students do not seem conceptually secure multiplying large numbers even with a calculator, you might encourage them to make smaller buildings so they have to deal only with smaller numbers.

Below is one example of a student's set of patterns for her buildings. She could have used some more of the leftover area to make a few other small buildings. Note that she has written how much room there is in a building (that is, how many cubes it will hold) in the center of each pattern.

When students submit their city plans at the end of the project, they will need to include four sheets like these, with their patterns drawn on the grids. You may want to prepare an overhead of this set of patterns (or something similar) to illustrate what a city plan looks like.

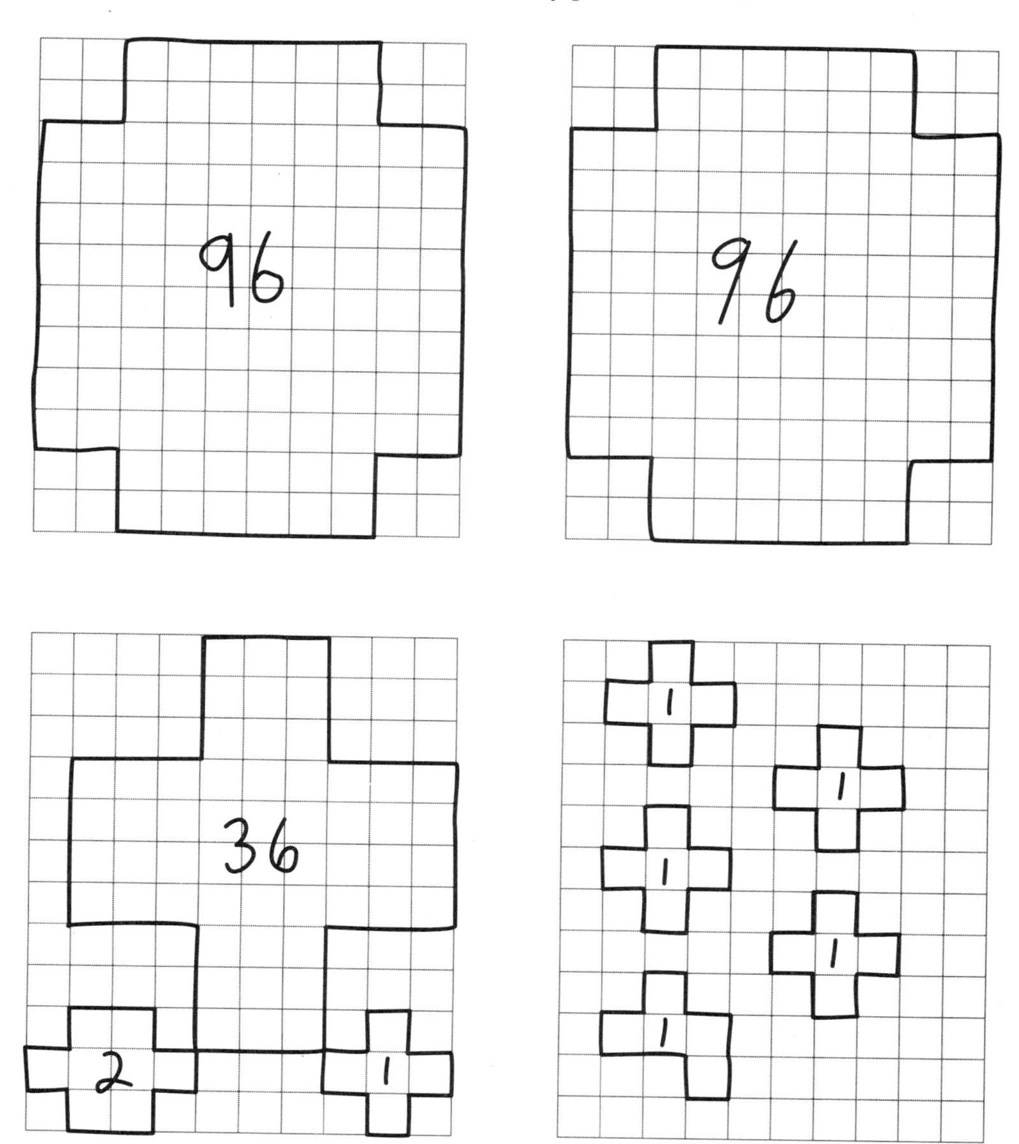

Once students have completed their individual plans, group them in pairs. Each pair should decide on one set of plans for a single city. This might be one of their two original plans, a combination of the two, or even a totally new set. Let students use graph paper and cubes to try other possibilities.

As students work together on their buildings, continue to circulate and discuss their work with them. Ask questions such as the following:

How many cubes fit in the building made by this pattern? How did you figure that out? Which pattern makes this building? How many cubes fit in your entire city?

When finding cube totals for the entire city, students often get different totals at different times. When this happens, encourage students to calculate the total yet again, this time being very careful and deliberate.

Take notes on the strategies students are using, problems they are experiencing, and the contributions of each student in a pair. Keep these notes to use in conjunction with your assessment of student work at the end of the project.

You will find that, for many students, this time will be the most productive of the entire *Exploring Solids and Boxes* unit. Because they become personally invested in making their city, they will be motivated to interconnect and become proficient with material in the previous investigations.

You may notice that some students' original city plans involve relatively little thought and creativity. But once these students start building and seeing their classmates' cities develop, given sufficient time and encouragement, they will generally be motivated to go back to the drawing board. They will start to discuss their design plans more carefully, including thinking about the size and shape of each building, and how patterns can be placed to minimize leftover graph paper. Their "ownership" of the city will motivate them to invest the effort required for success.

See the **Dialogue Box,** Thinking About Our Cities (p. 76), for some illustrations of how students' thinking about their cities evolved in one classroom.

Activity

Final Plans and Construction

When a pair of students has decided on a final set of buildings, they are ready to put together their final plans and build their cities.

Review what students need for their final submission (listed on p. 69). Remind them that they will need two identical sets of patterns for their buildings. They will cut out and tape together the patterns in one set to make buildings, then attach these to a sheet of construction paper to make their model of the city.

The other set is the city plan, which shows the four sheets of graph paper used, with each pattern drawn and labeled with the number of cubes that fit in the buildings (as shown on page 71). Remind students that they also need to compute the total number of cubes that will fit in their city.

All building patterns and calculations should be double-checked to be sure they are correct. Emphasize the usefulness of calculators in finding the number of cubes in buildings and for the whole city.

Finally, when everything has been double-checked, students should describe the design goals for their city on notebook paper. That is, why did they design their city the way they did?

❖ **Tip for the Linguistically Diverse Classroom** Students with limited English proficiency can draw their explanations of why they made their cities as they did.

This entire project will take three or four days. Students must be given plenty of time to work on their cities. At the start of each new day, review the task. Once students have completed their architectural models, they can color and decorate their buildings.

Assessment

Examining Students' City Plans

Collect the students' final plans to help you assess their progress. For this assessment, examine the notes you took on their planning strategies as they worked, as well as their final box patterns. Consider the following questions:

- Have students learned to make 2-D patterns that result in open rectangular boxes? What were their strategies for doing this? Did they use cube buildings to make their patterns, or did they use mental imagery? Are their boxes correct on the first try, or does it take several alterations to get them right?
- What strategies have students used to accurately determine the number of cubes that fit in their buildings? Do they need to fill up a building with cubes and then count the cubes? Do they have a "layering" strategy in which they figure out the number of cubes in one layer and then figure out how many layers are needed? If they employ the more sophisticated layering strategy, can they carry it out accurately? Can students predict how many cubes will fit in their box before they fold up their pattern, or do they need to see the 3-D shape in order to imagine how many cubes it holds?

- How have students determined the number of cubes in their entire city? Again, have they simply counted the cubes one by one? If so, do they have good strategies for double-checking their counting? Do they use effective addition or multiplication strategies to combine the number of cubes in all the buildings?
- How well have students explained their plans for making a city, and how well do their intentions fit the product? For example, students who say they want to use only a little land should have made buildings that are tall rather than flat. Have they understood this relationship?
- Have students experimented and taken risks? Have they tackled a variety of different kinds of structures, or have they restricted themselves to a few familiar building sizes?
- Did students learn from their work and modify their plans during the process? Did they learn from what they saw other students do and from their own attempts? Did they incorporate what they learned into revised plans for their cities?

Activity

Choosing Student Work to Save

As the unit ends, you may want to use one of the following options for creating a record of students' work on this unit:

- Students look back through their folders or notebooks and write about what they learned in this unit, what they remember most, and what was hard or easy for them. You might have students do this work during their writing time.
- Students select one or two pieces of their work as their best work, and you also choose one or two pieces of their work to be saved in a portfolio for the year. You might include students' work for the assessment An 18-Cube Box Pattern (p. 64), copies of their city plans or photos of their final cities from Investigation 5, and any other assessment tasks from this unit. Students can create a separate page with brief comments describing each piece of work.
- You may want to send a selection of work home for parents to see. Students write a cover letter, describing their work in this unit. This work should be returned if you are keeping a year-long portfolio of mathematics work for each student.

Sessions 1, 2, 3, and 4 Follow-Up

Extensions

Maximizing the Room Challenge As an added challenge, encourage students to try to maximize the amount of space in their city's buildings. That is, they should try to design a city whose buildings contain as many cubes as possible. Students will approach this task with different levels of sophistication. Some will use unsystematic trial and error. Others will be more reflective, trying to visualize boxes that hold the most cubes, or trying to figure out how to change buildings they have already made to increase the amount of space.

Great Connections This activity can easily and profitably be connected to one or more other school subjects.

- For economics, you might explain to students that in real life, companies are awarded contracts to build things (buildings, airplanes, cereal boxes, and so on) based on competition. They usually submit costs, specifications, drawings, and models to be evaluated. Students might find ways to assign "costs" to their buildings, perhaps based on the amount of paper used to construct them.
- To relate this project to social studies, emphasize the class discussion of the social and economic factors that are important in building a city. For instance, if the buildings are dwellings, even though large buildings maximize the room inside, the inhabitants might feel crowded. On the other hand, in big cities, land is very expensive, so builders make tall buildings with small bottoms that cover as little land as possible.
- You can integrate this project with art as students color and add detail to their buildings, also adding roads, trees, and local landmarks to the cities.
- To integrate the project with language arts (and social studies), one teacher asked students to write guidebooks for their cities.

We built a big city that has all the peopol's needs in it and it even has a city hall. We built it very cheeply so that homeless peopol could aford it because we care about them.

We built this city because the other cities didn't have enough space. We made the small buildings houses and shops. We made the big buildings office and apartment buildings.

Thinking About Our Cities

For the Building a City investigation, one pair of students, Maya and Khanh, had given minimal thought to their city at the beginning of the planning process. But once they started building their city and seeing other students' work, they decided to make some revisions. The teacher talked to them about their plans.

What made you decide to make more buildings?

Maya: Because we had extra space on our planning sheets.

Khanh: Then I decided to make a public pool.

Maya: And a restaurant!

How many cubes are in your city?

Khanh: 120.

How many did you have before you added the new buildings?

Khanh: 87. We added 33.

Another pair of students decided to increase the amount of cubes that their city would hold.

What can you do to make your city have more cubes?

Kate: Make more malls.

Midori: Make four malls.

How many cubes would that give?

Kate: [*Draws four 5-by-8-by-2 patterns*] 80 + 80 + 80 + 80 = 320 [*using a calculator*].

Midori: But the Building Code says that not all the buildings can be the same. Besides, we need some houses for people to live in.

Kate: So we could only have three malls.

Can you think of any other ways you could get more cubes in your city?

Midori: We could make skyscrapers. It wouldn't take up any more valuable room. But it would be more.

Kate: Our motel could have another story.

In both of these episodes, we see students thinking about their buildings from both a numerical and a social utility perspective. Students are using mathematics to help them think about the real world, an essential part of making mathematics useful to them.

Quick Images

Basic Activity

Students are briefly shown a picture of a geometric object. Depending on the kind of object, they either build it or draw it by developing and inspecting a mental image of it.

For each type of problem—2-D shapes and cube images—students must find meaningful ways to construct and interpret a mental image of the object. They might see it as a whole ("it looks like a box, three cubes long and two cubes high"), or decompose it into memorable parts ("it looks like four triangles, right side up, then upside down, then right side up, then upside down"). Their work focuses on:

- organizing and analyzing visual images
- developing concepts and language needed to reflect on and communicate about spatial relationships
- using geometric vocabulary to describe shapes and patterns
- using number relationships to describe patterns

Materials

- Overhead projector
- Overhead transparencies of the geometric figures you will use as images for the session; we provide two pages of transparency masters to get you started. To use the images, first make a transparency, then cut out the separate figures and keep them in an envelope. Include the numbers beside the figures; they will help you properly orient the figures on the overhead.
- Pencil and paper (if you are using 2-D images)
- Interlocking cubes: 15–20 per student (if you are using figures of cube buildings)

Procedure

Step 1. Flash an image for 3 seconds. You might show a picture of a geometric drawing or a cube building.

It's important to keep the picture up for as close to 3 seconds as possible. If you show the picture too long, students will work from the picture rather than their image of it; if you show it too briefly, they will not have time to form a mental image. Suggest to students that they study the figure carefully while it is visible, then try to draw or build it from their mental image.

Step 2. Students draw or build what they saw. Give students a few minutes with their pencil and paper or the cubes to try to construct or draw a figure based on the mental image they have formed. After you see that most students' activity has stopped, go on to step 3.

Step 3. Flash the image again, for revision. After showing the image for another 3 seconds, students revise their building or drawing, based on this second view.

It is essential to provide enough time between the first and second flashes for most students to complete their attempts at building or drawing. While they may not have completed their figure, they should have done all they can until they see the picture on the screen again.

When student activity subsides again, show the picture a third time. This time leave it visible, so that all students can complete or revise their solutions.

Step 4. Students describe how they saw the drawing as they looked at it on successive "flashes."

Variations

We provide transparency masters for two types of Quick Images: geometric designs and cube buildings. You can supplement these with your own examples or make up other types.

Continued on next page

Quick Image Cubes Each student should have a supply of 15-20 cubes and be seated facing the overhead screen. Show a picture from the Quick Image Cubes transparency; note that the transparency should be cut apart so that only one picture shows at a time. Proceed as described. You may be able to show two or three Quick Images in a ten-minute session. Some teachers have found it valuable to repeat, on successive days, Quick Images that were previously presented.

Quick Image 2-D Geometric Designs Use the Quick Image 2-D Geometric Designs transparency. Follow the same procedure as for the cubes, but have students draw the images they see.

When students talk about what they saw in successive flashes, many students will say things like "I saw three triangles in a row." You might suggest this strategy for students having difficulty: "Each design is made from familiar geometric shapes. Find these shapes and try to figure out how they are put together."

As students describe their figures, you can introduce correct terms for them. As you use them naturally as part of the discussion, students will begin to use and recognize them.

Related Homework Options

- **Creating Quick Images** Students can make up their own Quick Images to challenge the rest of the class. Talk with students about keeping these reasonable—challenging, but not overwhelming. If they are too complex and difficult, other students will just become frustrated.
- **Family Quick Images** You can also send images home for students to try with their families. Instead of using the overhead projector, they can simply show a picture for a few seconds; cover it up while members of the family try to draw it; then show it again, and so forth. Other members of the family may also be interested in creating images for the student to try.

What Is Likely?

Basic Activity

Students make judgments about drawing objects of two different colors from a clear container. They first decide whether it's likely that they'll get more of one color or the other. Then they draw out 10 objects, one at a time, recording the color of each and replacing that object before picking the next one. Students then discuss whether what they expected to happen did happen. They repeat the activity with another sample of 10 objects.

What Is Likely? involves students in thinking about ratio and proportion, and in considering the likelihood of the occurrence of a particular event. Ideas about probability are notoriously difficult for children and adults. In the early and middle elementary grades, we simply want students to examine familiar events in order to judge how likely or unlikely they are. In this activity, students' work focuses on:

- visualizing the ratio of two colors in a collection
- making predictions and comparing predictions with outcomes
- exploring the relationship between a sample and the group of objects from which it comes

Materials

- A clear container, such as a fishbowl or large glass or clear plastic jar
- Objects that are all very similar in size and shape, but come in two colors (wooden cubes, beans, beads, marbles)

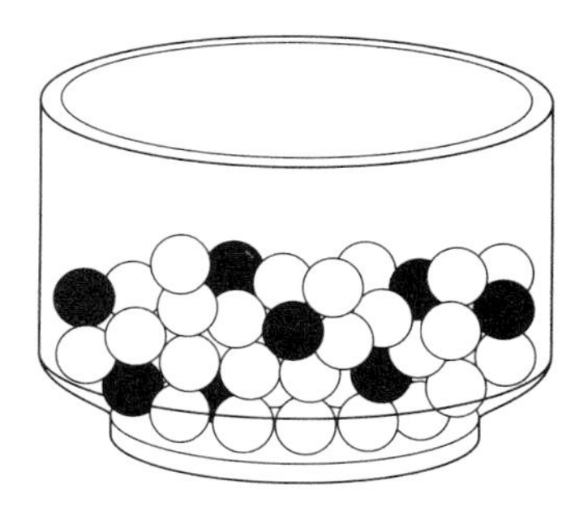

Procedure

Step 1. Fill the container with two colors of cubes, beads, or beans. When you first use this activity, put much more of one color into the container. For example, out of every 10 cubes you put in the container, you might use 9 yellow and 1 red. Thus, if you used 40 cubes, 36 would be yellow and 4 would be red. Mix these well inside the container. Continue to use these markedly different proportions for a while.

Step 2. Students predict which color they will get the most of if they draw 10 objects out of the container. Carry the container around the room so that all students can get a good look at its contents. Then ask students to make their predictions. "What is likely to happen if we pull out 10 objects? Will we get more yellows or more reds? Will we get a lot more of one color than the other?"

Step 3. Students draw 10 objects from the container, replacing after each draw. Ask a student, with eyes closed, to draw out one object. Record its color on the board before the student puts the object back. Ask nine more students to pick an object, then replace it after you have recorded its color. Record colors using a table and tallies.

RED ~~////~~ ///

YELLOW //

Step 4. Discuss what happened. "Is this about what you expected? Why or why not?" Even if you have a 9:1 ratio of the two colors, you won't always draw out a sample that is exactly 9 of one color and 1 of the other. Eight red and 2 yellow or 10 red and 0 yellow would also be likely samples. Ask students whether what they got is likely or unlikely, given what they can see in the container. What would be *unlikely,* or surprising? (Of course, surprises can happen, too—just not very often!)

Continued on next page

Step 5. Try it again. Students will probably want to try drawing another 10 objects to see what happens. "Do you still think it's likely that we'll get mostly reds again? Why? About how many do you think we'll get?" Draw objects, tally their colors, and discuss in the same way.

Variations

Different Color Mixes Try a 3:1 ratio—3 of one color for every 1 of the other color. Also try an equal amount of the two colors.

Different Objects Try two colors of a different kind of object. Does a change like this affect the outcome?

The Whole Class Picks See what happens when each student in the class draws (and puts back) one object. Before you start, ask, "If all 28 of us pick an object, about how many reds do you think we'll get? Is it more likely you'll pick a red or a yellow? A *little* more likely or *a lot* more likely?"

Students Fill the Container Ask students to help you decide what proportions of each color to put in the container. Set a goal, for example:

- How can we fill the container so that's it's very likely we'll get mostly reds when we draw 10?
- How can we fill the container so that it's unlikely we'll get more than one red?
- How can we fill the container so that we'll get close to the same number of reds and yellows when we draw 10?

After students decide how to fill the container, draw objects, as in the basic activity, to see if their prediction works.

Three Colors Put an equal number of two colors (red and yellow) in the container, and mix in many more or many fewer of a third color (blue). "If 10 people pick, about how many of each color do you think we will get? Do you think we'll get the same number of red and yellow, or do you think we will get more of one of them?"

The following activities will help ensure that this unit is comprehensible to students who are acquiring English as a second language. The suggested approach is based on *The Natural Approach: Language Acquisition in the Classroom* by Stephen D. Krashen and Tracy D. Terrell (Alemany Press, 1983). The intent is for second-language learners to acquire new vocabulary in an active, meaningful context.

Note that *acquiring* a word is different from *learning* a word. Depending on their level of proficiency, students may be able to comprehend a word upon hearing it during an investigation, without being able to say it. Other students may be able to use the word orally, but not read or write it. The goal is to help students naturally acquire targeted vocabulary at their present level of proficiency.

We suggest using these activities just before the related investigations. The activities can also be led by English-proficient students.

Investigation 1

point

1. As you show students a sharpened pencil, pen, nail, and pyramid, identify the place where each comes to a *point*.
2. Ask several students to find two objects in the room—one with a point and one without a point.

edge

1. Show students the *edges* of a piece of paper. Have them count the edges.
2. Show the edges of several other things in the classroom, such as a table, a cabinet, a box, a shelf. Choose edges that are fairly sharp and well defined. Then have students show more edges, perhaps of things at their own desks.
3. Ask students to clap when you touch an edge and snap their fingers when you touch a surface that is not an edge. Again, choose well-defined edges, and contrast them with tabletops, the surfaces of a globe or cylindrical vase, and so on.

Investigation 3

overlap

1. Arrange three pieces of paper side by side on a table so that they are close together but not overlapping. As you point to the line between each pair, explain that none of the papers are *overlapping*.
2. Rearrange the papers so they overlap. As you do so, show where each paper is *overlapping* another.
3. Have each student take two pieces of paper. Challenge students' comprehension by creating action commands:

   Make the two papers overlap each other.
   Show me where they overlap.
   Move the papers so they do not overlap.

Blackline Masters

_________________ , 19 ____

Dear Family,

For the next few weeks, we will be doing a mathematics unit called *Exploring Solids and Boxes.* The goal of this unit is for the children to become comfortable with the characteristics of basic geometric shapes, such as pyramids, cubes, and prisms.

Learning about three-dimensional (3-D) shapes is an often-overlooked part of mathematics. This is surprising, because the ability to reason about 3-D shapes is a very practical mathematical skill. Jobs ranging from architecture to fashion design to construction work require that people know how to think about and manipulate shapes. Many jobs also require workers to read two-dimensional plans for geometric shapes, and to translate the plans into three dimensions. These are just the kinds of skills that your child will be working on in this unit.

In this module, the class will be building common 2-D and 3-D shapes. They will make open boxes out of paper and fill the boxes with small cubes. This gives them a chance to develop their ideas of how much space is actually in a box. As a final project, students design and build a city out of paper boxes and figure out how much room is in their buildings.

For homework, your child might be asked to find objects that have the same shape as those we are studying in class. Your child may also be asked to build some shapes with straws and clay (or toothpicks and small marshmallows), and to draw paper patterns that fold to make different geometric shapes.

You can help your child by showing an interest in and talking about what the class is doing. Another way you can help throughout this unit, and throughout the year, is to encourage your child to work with building toys and materials that you have around the house. Third graders are not too old to play with blocks—nor are their parents! This is play that serves an important learning need.

Sincerely,

Name Date

Identifying Geometric Shapes in the Real World (page 1 of 3)

Geometric Shapes	Real-World Objects

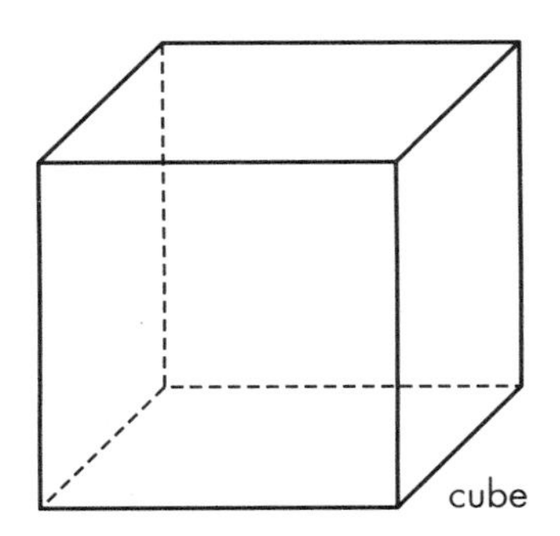

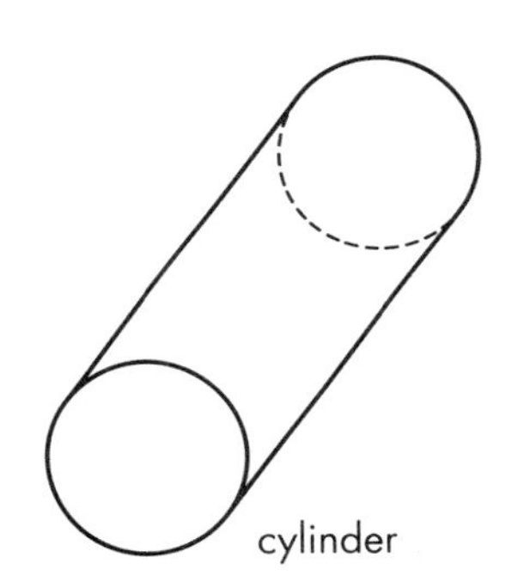

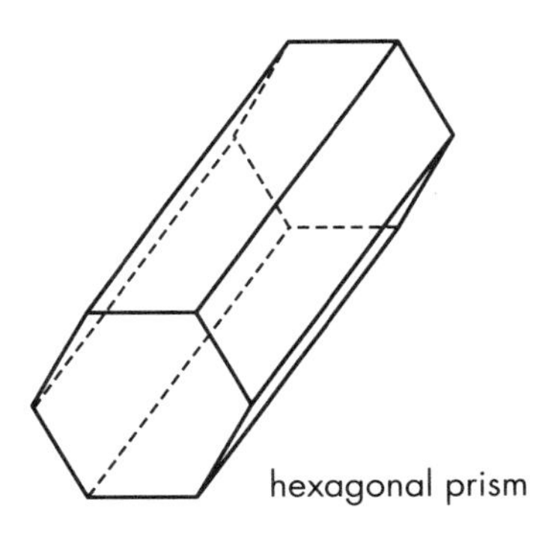

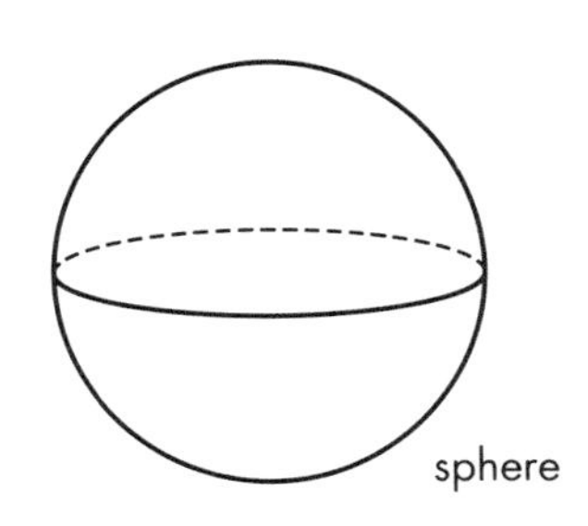

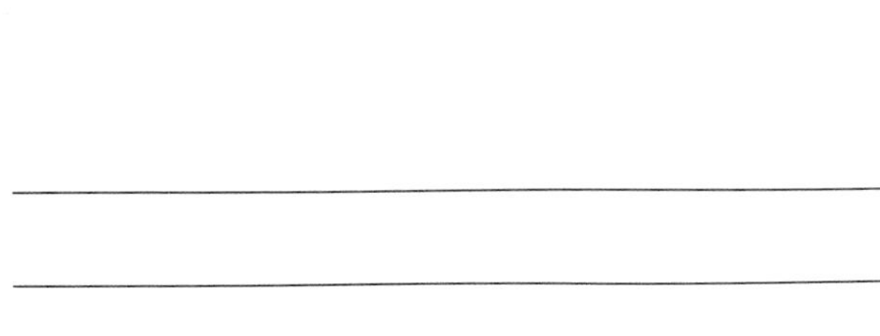

Identifying Geometric Shapes in the Real World (page 2 of 3)

Geometric Shapes | Real-World Objects

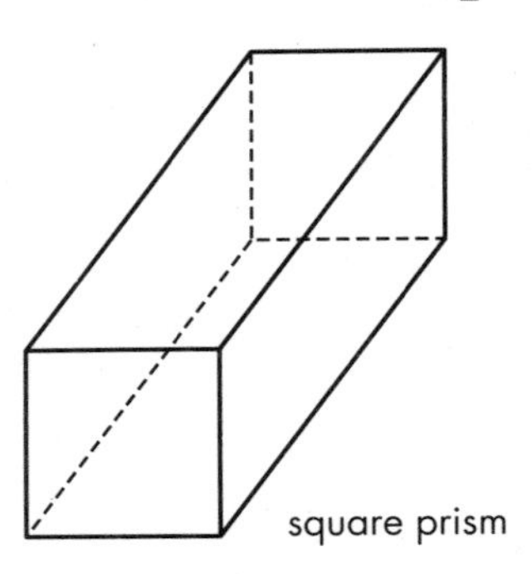

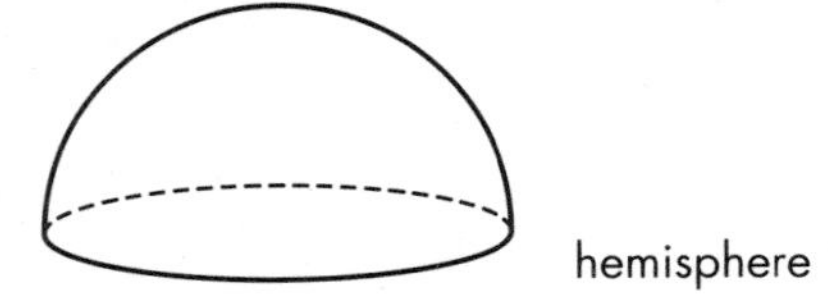

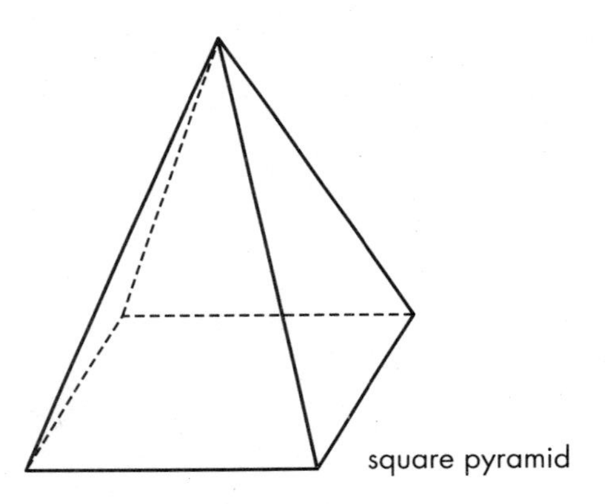

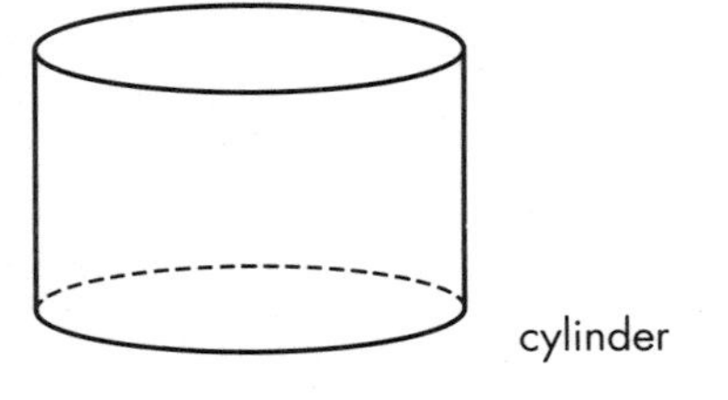

Identifying Geometric Shapes in the Real World (page 3 of 3)

Geometric Shapes	Real-World Objects

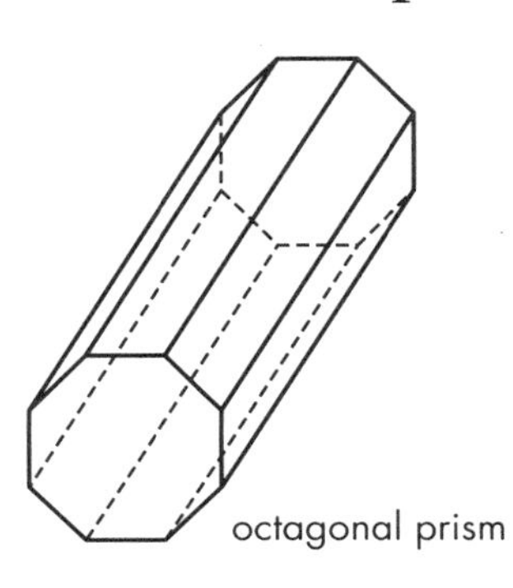
octagonal prism

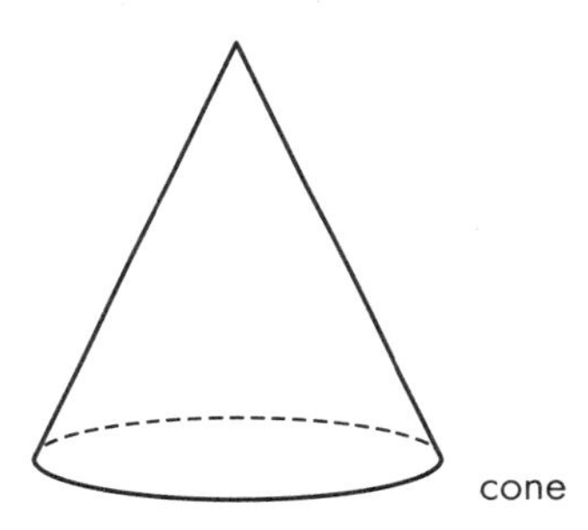
cone

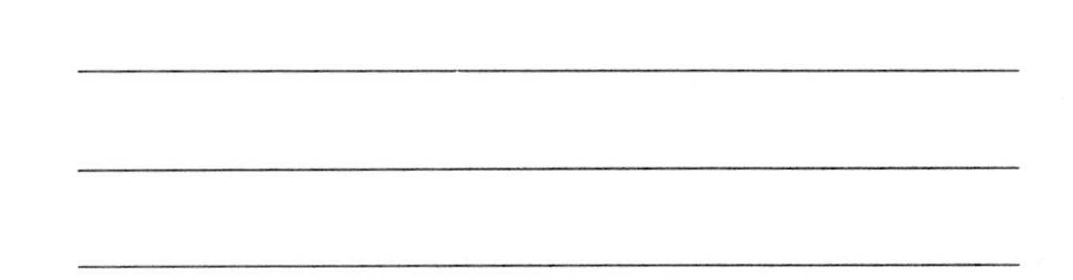

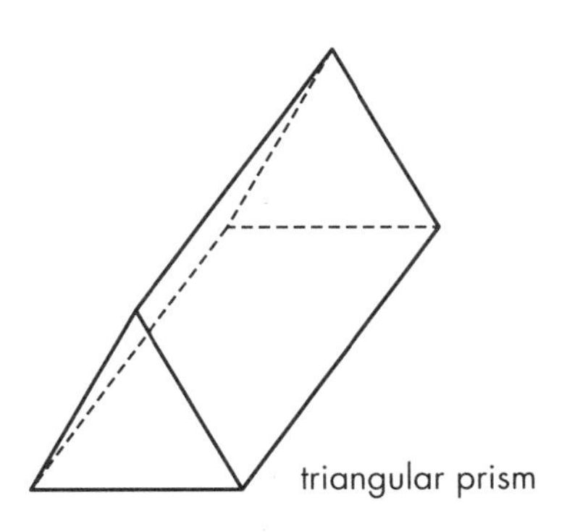
triangular prism

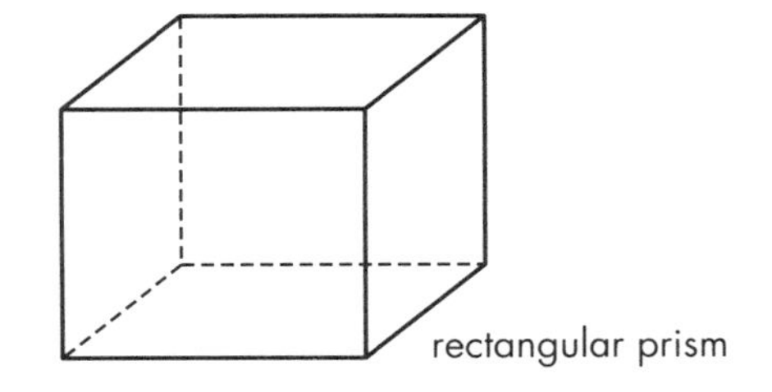
rectangular prism

Name Date

Student Sheet 2

Building Triangles

Build the triangles described below. Draw a picture of each triangle you build. Label the length of each stick you use.

1. Make a triangle with all sides the same length.	2. Make a triangle with all sides different lengths.
3. Can you find three sticks that will **not** make a triangle? If you can, draw them. Label them to show their lengths. Why won't these sticks make a triangle?	

CHALLENGE

How many different triangles can you build using one 5-inch stick, one 6-inch stick, and one 8-inch stick? Draw them on the back of this sheet.

Name Date

Student Sheet 3

Building Squares

1. Build two different squares. Draw a picture of each square. Label the length of each side.

2. Can you build a square with one 6-inch stick, one 5-inch stick, one 4-inch stick, and one 3-inch stick? Why or why not?

Name Date

Student Sheet 4

Building Rectangles

1. Build three different rectangles. Draw a picture of each rectangle. Label the lengths of the sides.

2. Can you build a rectangle with one 5-inch stick, one 6-inch stick, and two 3-inch sticks? Why or why not?

CHALLENGE

How many different four-sided polygons can you build using one 8-inch stick, one 6-inch stick, one 5-inch stick, and one 4-inch stick? Draw some of them on the back of this sheet. Be sure to label the lengths of the sides.

Name Date

Student Sheet 5

Geometric Solids

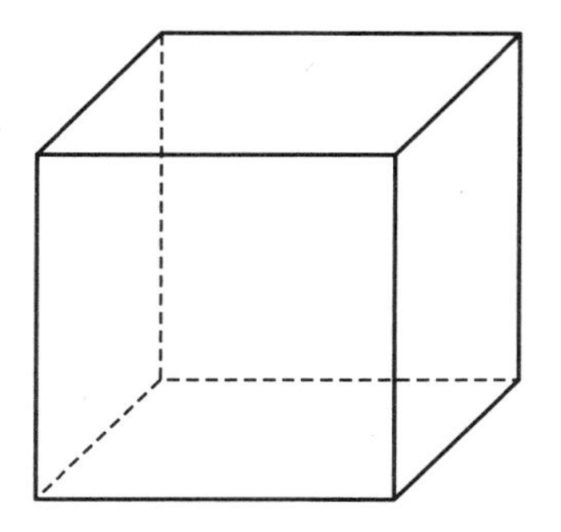
cube

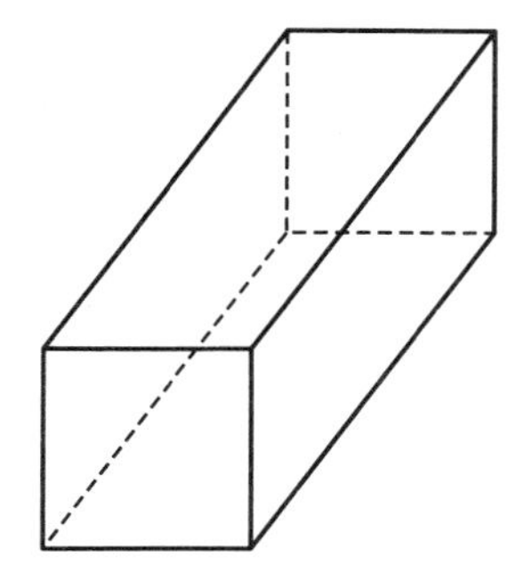
square prism

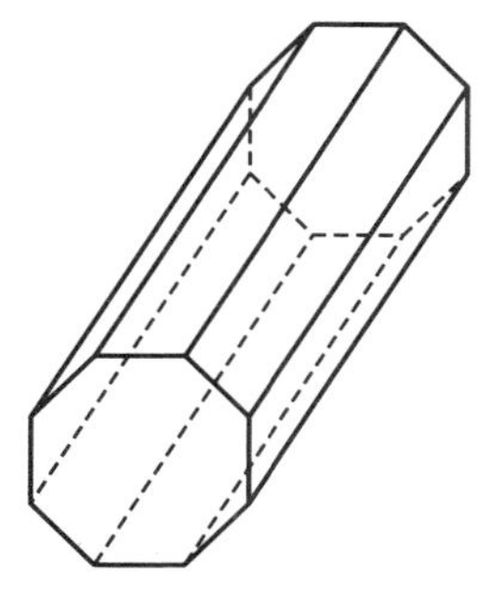
octagonal prism

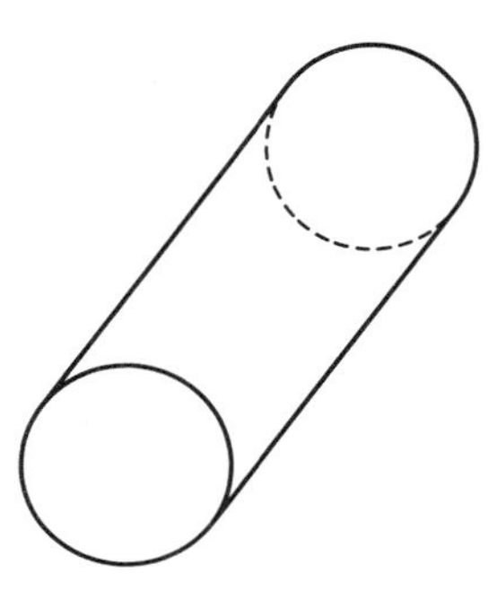
cylinder

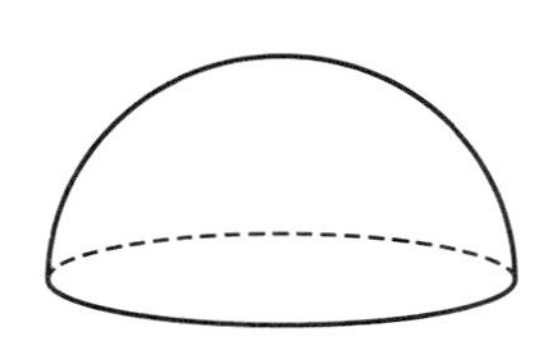
hemisphere

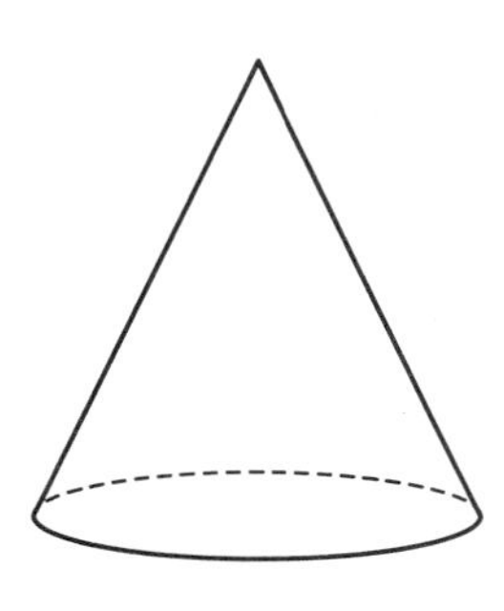
cone

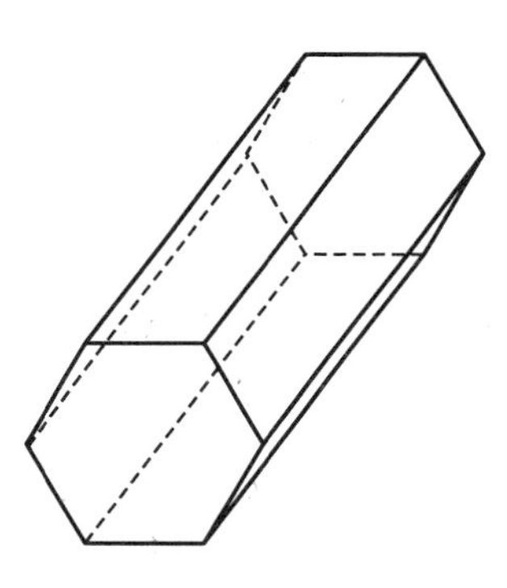
hexagonal prism

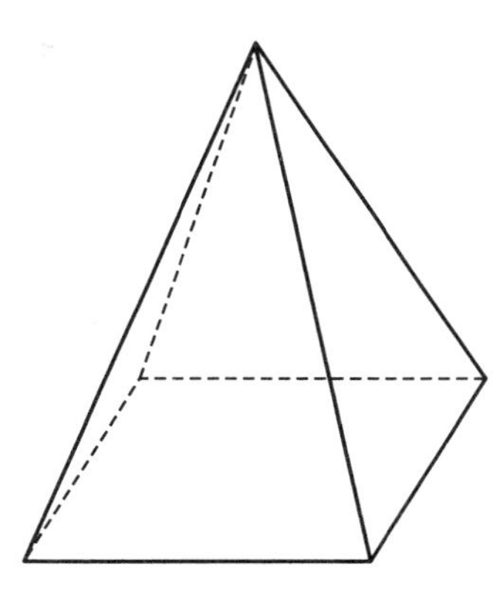
square pyramid

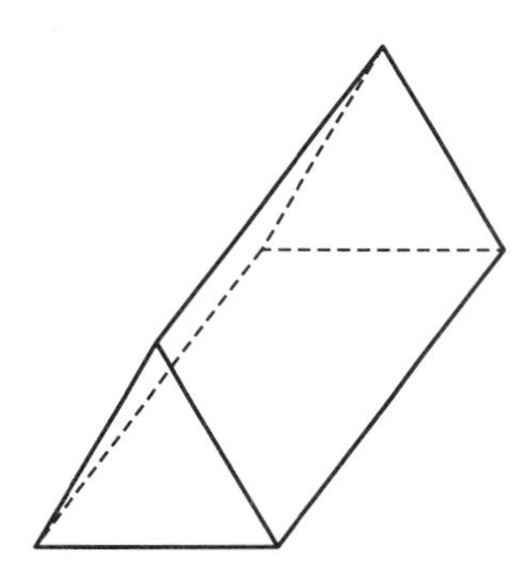
triangular prism

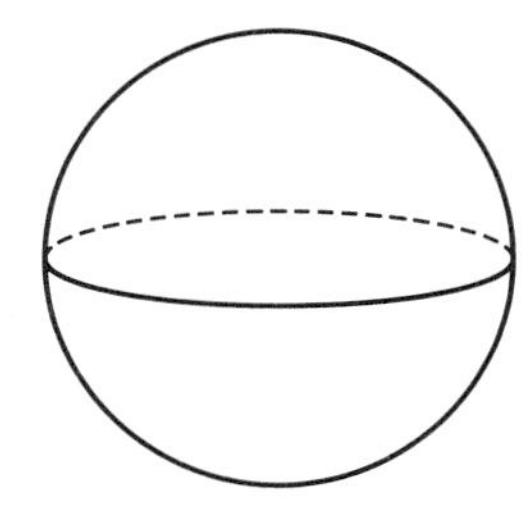
sphere

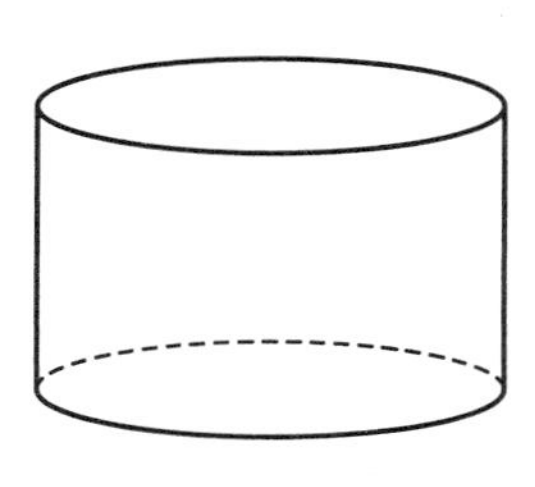
cylinder

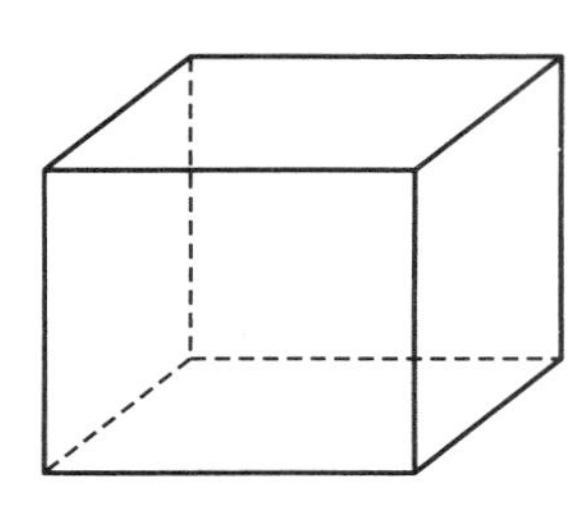
rectangular prism

Name Date

Student Sheet 6

Patterns for Cube Boxes

Which of these patterns do you think can be used to make an open box for a cube? Write YES by the letter if you think a pattern will make a box. Write NO if you think it won't make a box. Cut out the patterns to check your predictions.

Predictions

A. _____ B. _____ C. _____ D. _____ E. _____

A

B

C

D

E

Name Date

Student Sheet 7

More Patterns

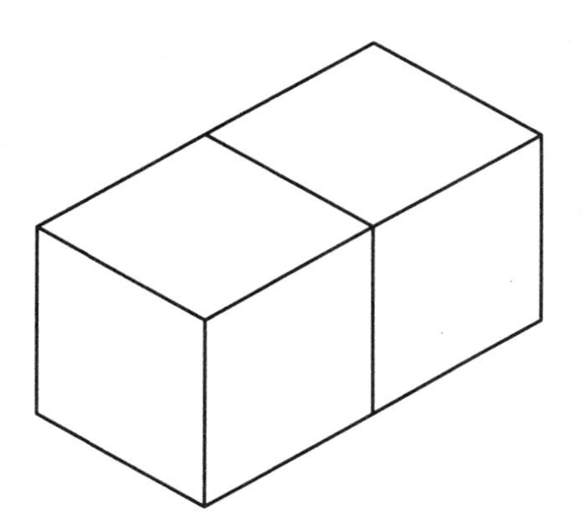

1. Make a pattern for an open box that holds two 1-inch cubes. Use graph paper. Find as many patterns as you can.

2. Use triangle grid paper to make patterns for a four-sided pyramid. All of its sides should look like the triangle below.

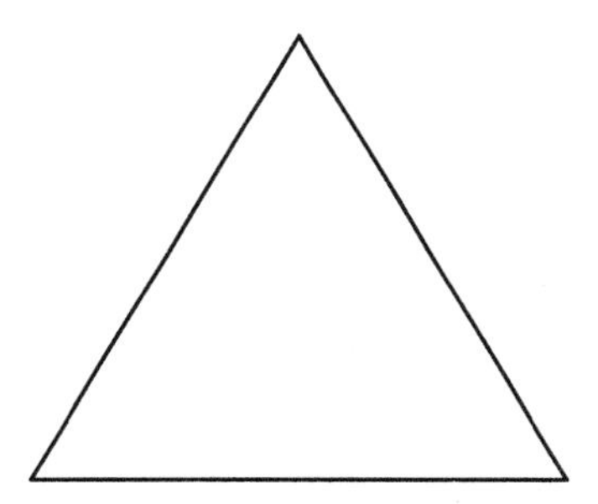

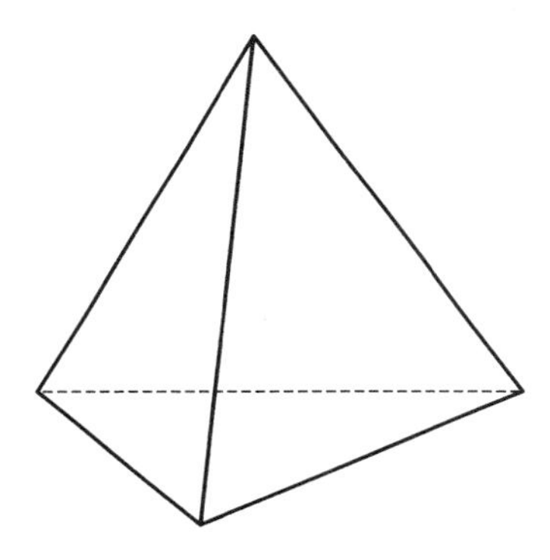

Your pyramid can be any size. Find as many patterns as you can.

four-sided pyramid

Name Date

How Many Cubes? (page 1 of 4)

Pattern A makes a rectangular box without a top.
How many cubes do you predict will fit in the box? ________

To check your answer, cut out the pattern, build the box, and fill the box with cubes.

A

How Many Cubes? (page 2 of 4)

Pattern B makes a rectangular box without a top.
How many cubes do you predict will fit in the box? ________

To check your answer, cut out the pattern, build the box, and fill the box with cubes.

B

How Many Cubes? (page 3 of 4)

Pattern C makes a rectangular box without a top.
How many cubes do you predict will fit in the box? ________

To check your answer, cut out the pattern, build the box, and fill the box with cubes.

C

How Many Cubes? (page 4 of 4)

Pattern D makes a rectangular box without a top.
How many cubes do you predict will fit in the box? ________

To check your answer, cut out the pattern, build the box, and fill the box with cubes.

D

CHALLENGE
Make some patterns for boxes that contain exactly 15 cubes.

Name Date

Making Boxes from the Bottom Up (page 1 of 3)

1. The dark squares make the bottom of a rectangular box that contains exactly 6 cubes. The box has no top. Draw the sides to finish the pattern for the box.

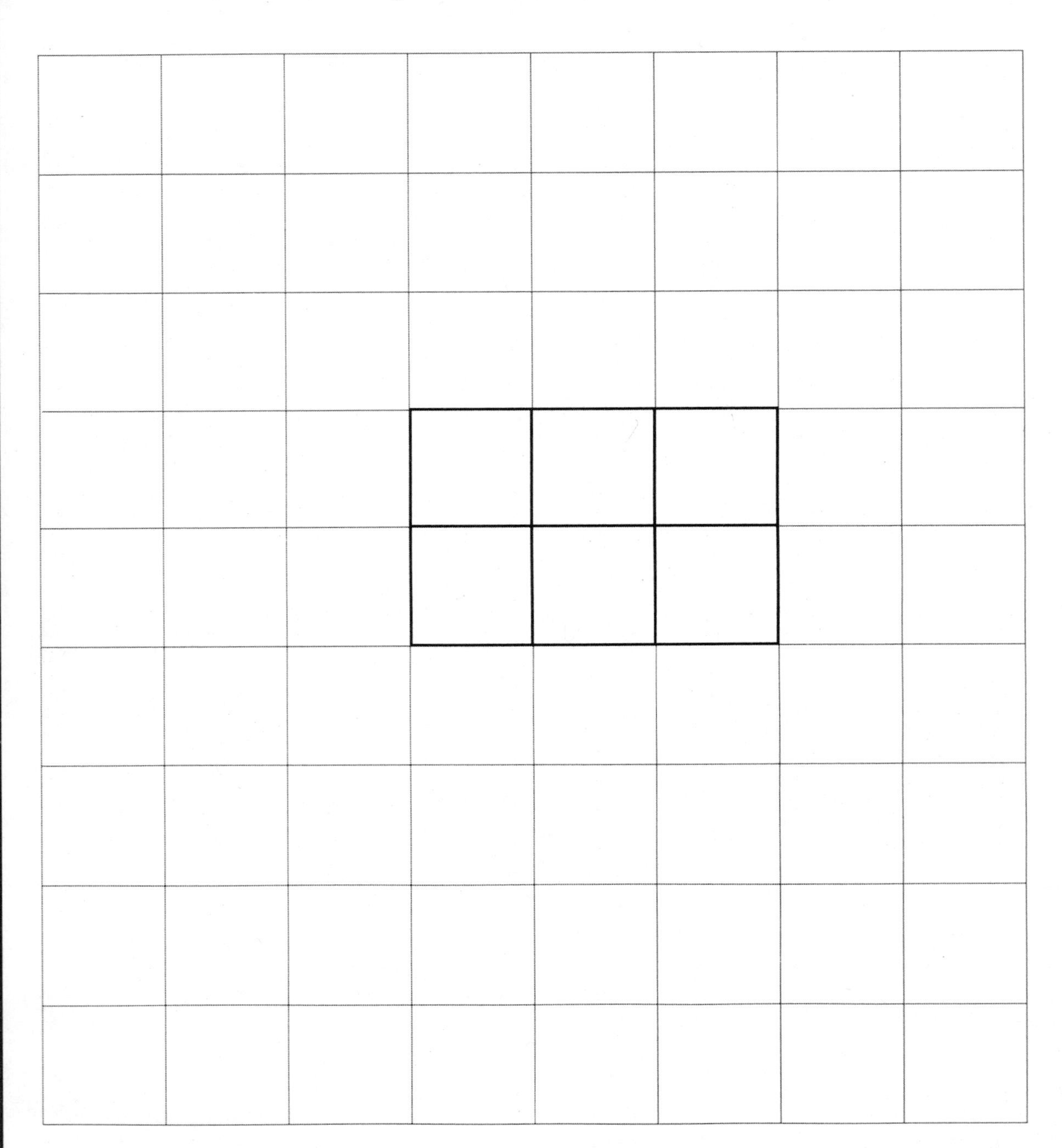

Making Boxes from the Bottom Up (page 2 of 3)

2. The dark squares make the bottom of a rectangular box that contains exactly 9 cubes. The box has no top. Draw the sides to finish the pattern for the box.

Making Boxes from the Bottom Up (page 3 of 3)

3. The dark squares make the bottom of a rectangular box that contains exactly 8 cubes. The box has no top. Draw the sides to finish the pattern for the box.

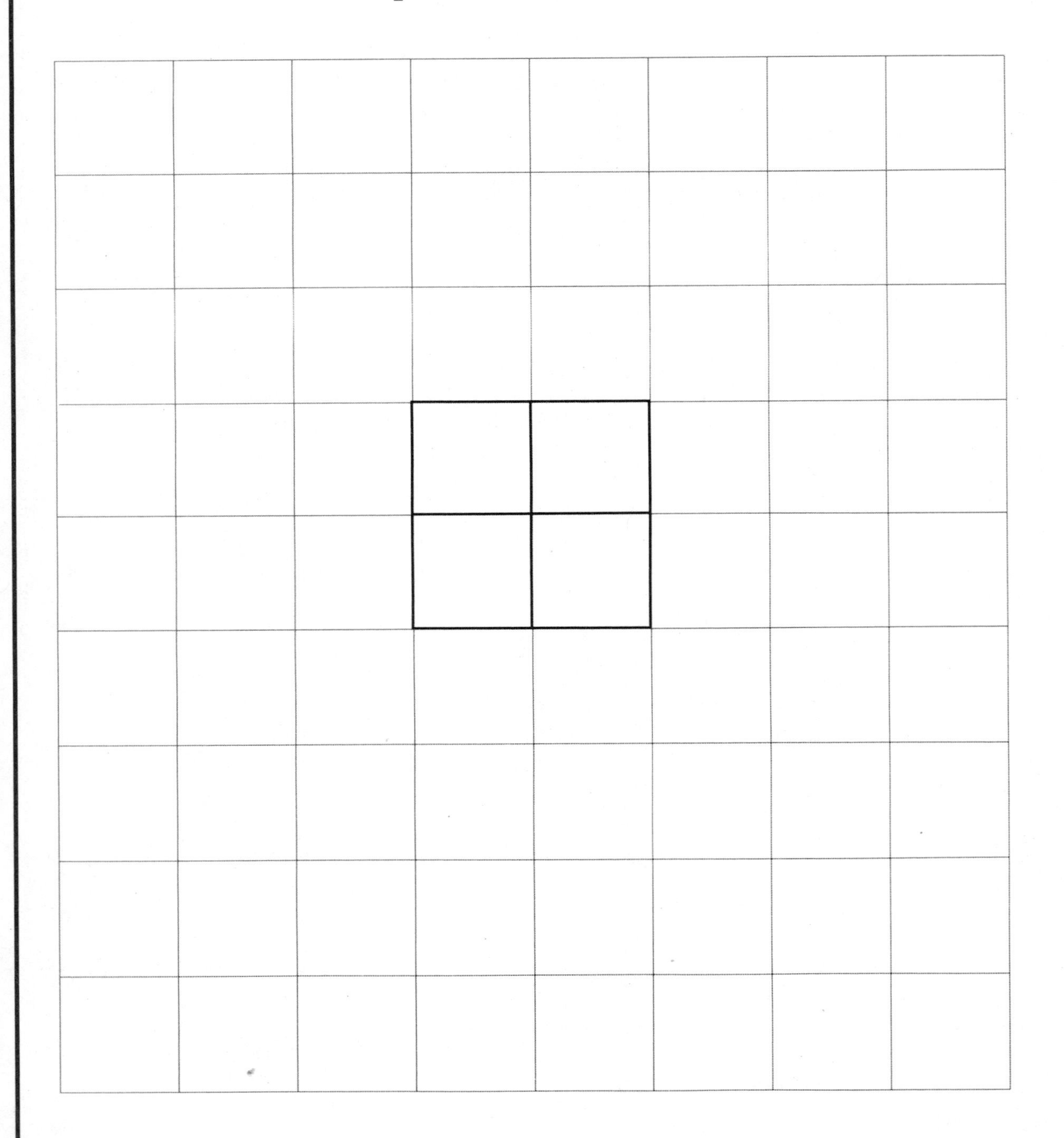

Name Date

Student Sheet 10

An 18-Cube Box Pattern

The dark squares make the bottom of a rectangular box that contains exactly 18 cubes. The box has no top. Draw the sides to finish the pattern for the box.

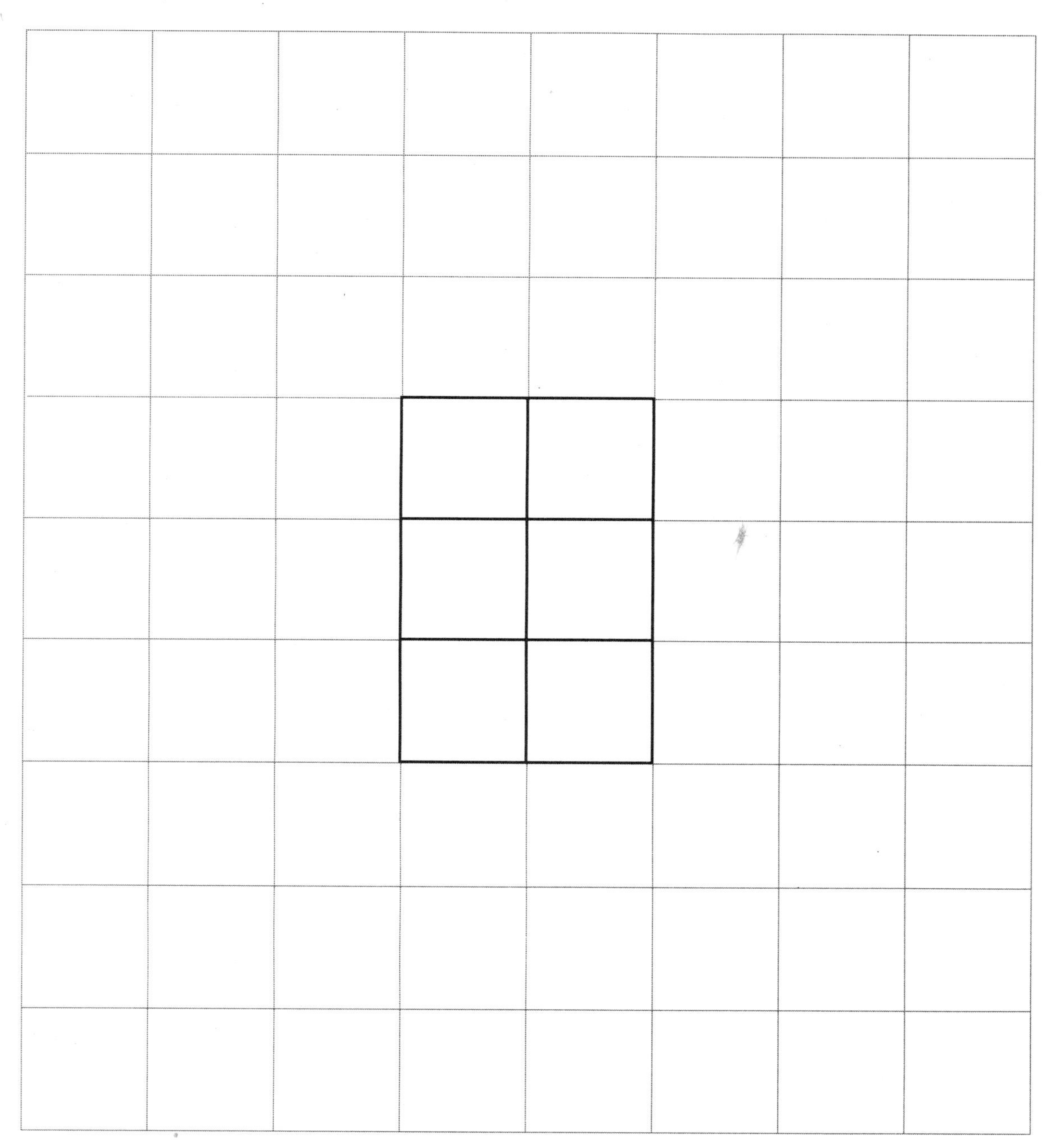

How did you figure out your pattern for this box? Write your answer on the back of this sheet.

Building Polyhedra

1. Build a polyhedron that has exactly 6 square faces.

2. Build a polyhedron that has exactly 1 square face and 4 triangular faces.

3. Build a polyhedron that has exactly 3 rectangular faces and 2 triangular faces.

4. Build a polyhedron that has exactly 8 corners and 6 faces.

5. Build a polyhedron that has exactly 5 corners and 5 faces.

6. How many differently shaped polyhedra can you make that have exactly 12 edges?

CHALLENGE
Make a polyhedron that has exactly 6 edges and 4 triangular faces.

Building Kit Length Guide

Sticks

2-inch

3-inch

4-inch

5-inch

6-inch

8-inch

Demonstration Box Pattern

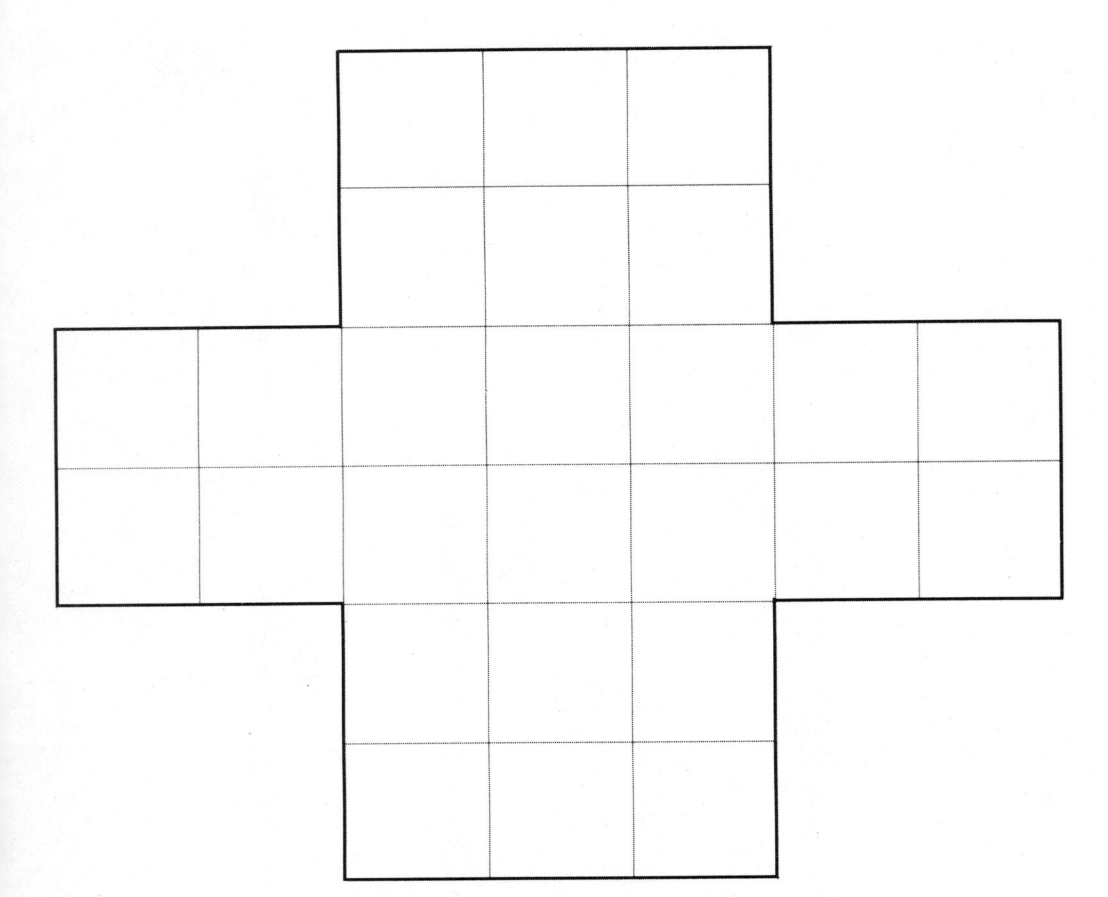

Discussion Box Pattern

TRIANGLE PAPER

Quick Image Geometric Designs

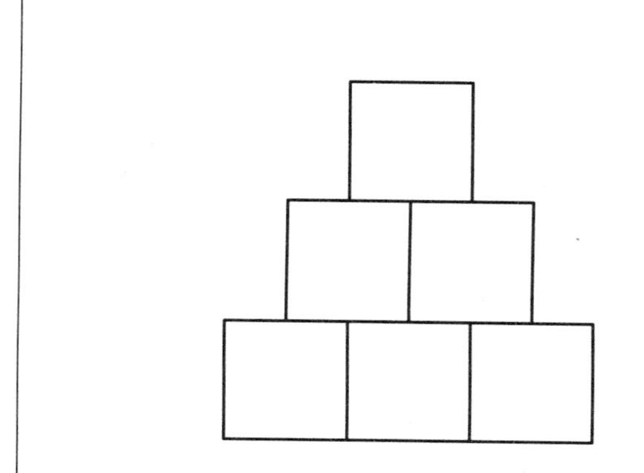

1.

2.

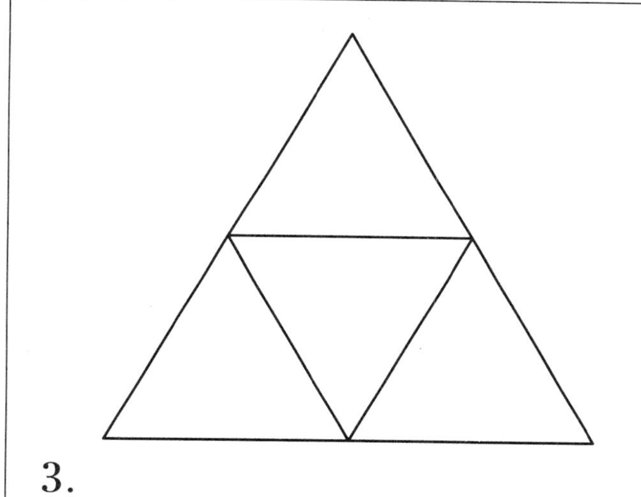

3.

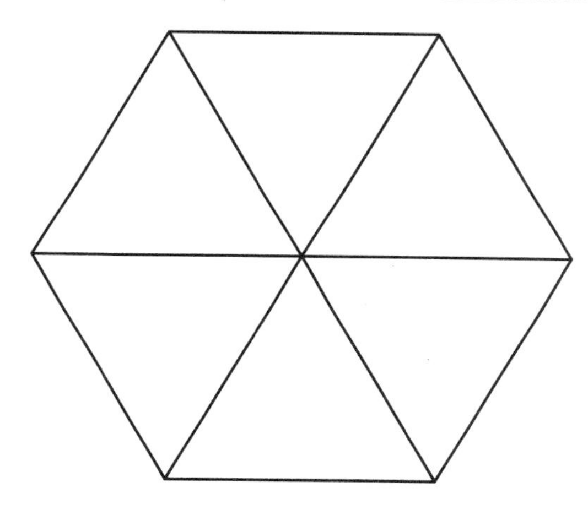

4.

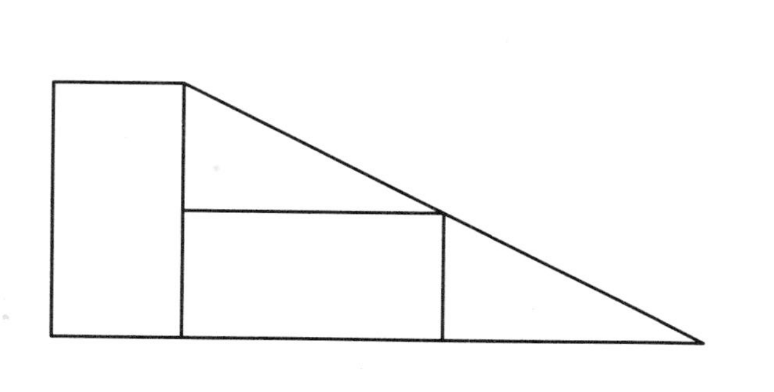

5.

6.

Quick Image Cubes

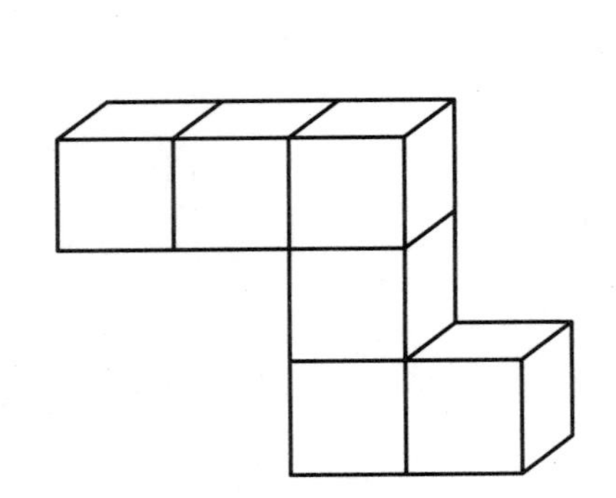

1.

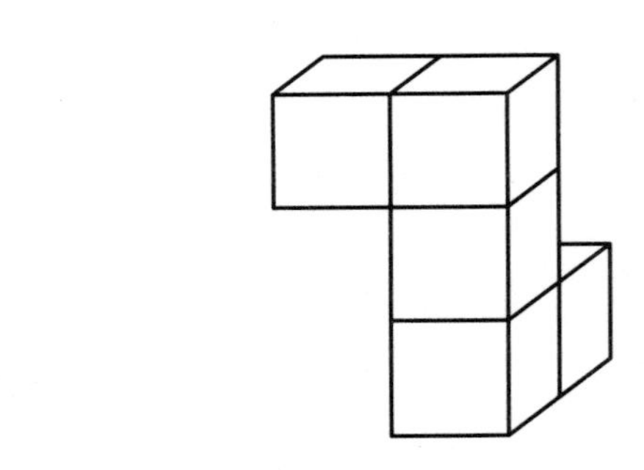

2.

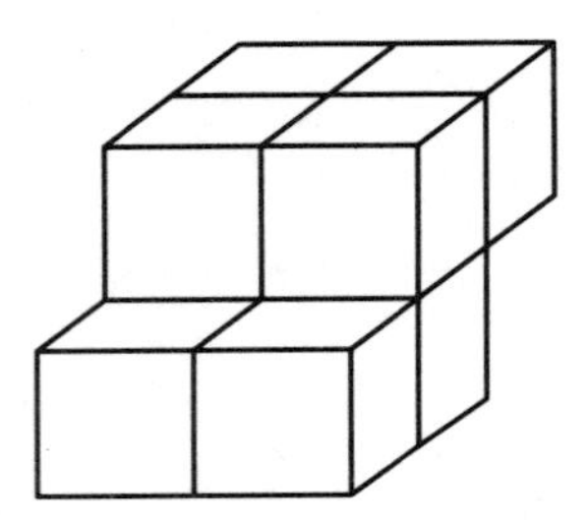

3.

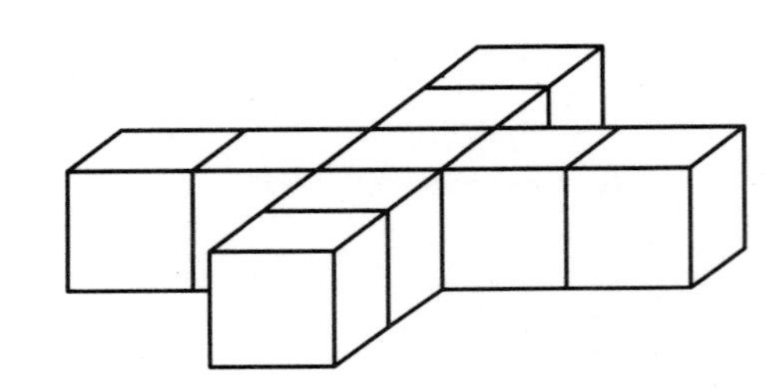

4.

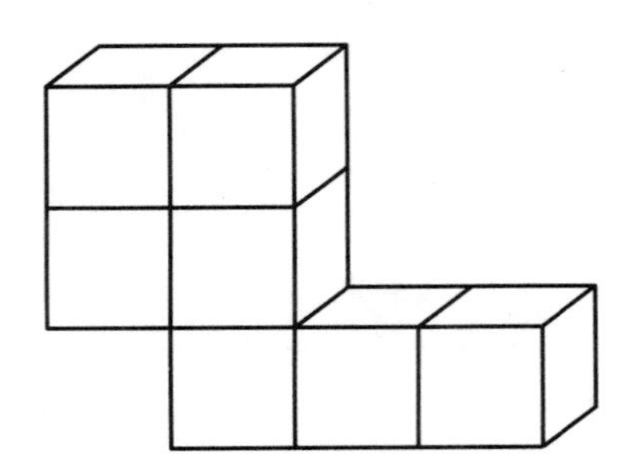

5.

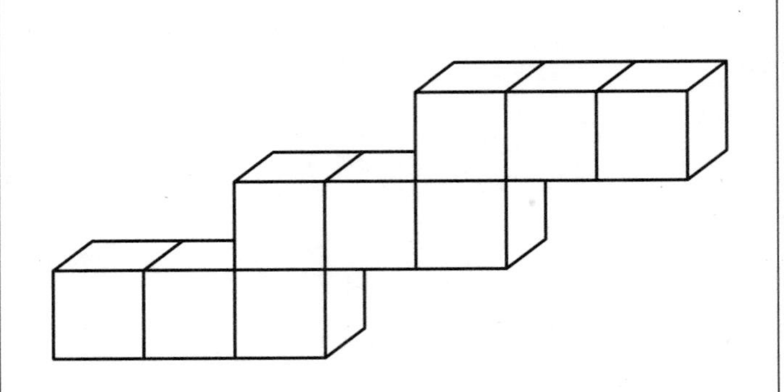

6.